Risk-Taking, Pre
and Design

Risk-Taking, Prevention, and Design

A Cross-Fertilization Approach

Edited by

Guy André Boy
Edwige Quillerou

CRC Press
Taylor & Francis Group
Boca Raton London New York

CRC Press is an imprint of the
Taylor & Francis Group, an **informa** business

First edition published 2023
by CRC Press
6000 Broken Sound Parkway NW, Suite 300, Boca Raton, FL 33487-2742
and by CRC Press

4 Park Square, Milton Park, Abingdon, Oxon, OX14 4RN

CRC Press is an imprint of Taylor & Francis Group, LLC

ISBN: 9781032118000 (hbk)
ISBN: 9781032118031 (pbk)
ISBN: 9781003221609 (ebk)

DOI: 10.1201/9781003221609

Typeset in Sabon
by KnowledgeWorks Global Ltd.

Contents

Preface

We boldly propose to engage this work as a manifesto of a scientific and societal transdisciplinary project for sociotechnical design projects. Coming from very different backgrounds, we have chosen to engage in a collaborative experience writing together and assembling various contributions from a large variety of domains. This exercise targeted a cross-fertilized integration of various kinds of expertise in engineering sciences with human and social sciences to produce HSI projects.

The collaboration between us, Guy André Boy and Edwige Quillerou, started from their collaborative work on human-centered design (HCD) projects and a reflection around the evolution of engineering professions. We use this collaboration to develop rationalize new ways of associating engineering design together with health and safety principles and issues. Consequently, the resulting human-centered vision opens a growing transdisciplinary approach, HSI.

This book provides an integrating approach that associates prevention and design. For a long time, engineering design and safety engineering have been two separate fields. HCD associated with Systems Engineering, where modeling and human-in-the-loop simulation are extensively used, provides a framework for this kind of integration. Indeed, we cannot dissociate technological design from organizational and human factors. Anytime new technology is developed and used, new jobs are created either in an evolutionary way or involving radical changes. This book addresses what engineering design and risk prevention mean though concrete examples, and how they can be associated for the development of life-critical systems, that is, considering workers' health and safety issues together with creativity at design time, and incorporating them into solutions that should be flexibly modifiable during the whole life cycle of systems.

Our fundamental objective is to create a framework for collaboration of engineering designers, systems engineers and work psychologists as well as other professionals including sociologists, ergonomists and management specialists. It is also important that this topic does not stay at a national level, but investigate, and at least discuss, possible commonalities and differences across cultures. Indeed, standardization cannot be developed context-free, and should provide enough flexibility for modification when required at any point in time.

Reliability and availability at work are essential. Rigid procedural methods developed in large organizations rapidly end up being inoperative, and flexibility becomes a real requirement especially when things go wrong. Flexibility is about autonomy that cannot be developed without appropriate authority sharing, competence, collaboration and trust. Even when all precautions have been taken, there is always a time when people have to take action, and of course take risk.

This book aims at providing means for a transformation of current sociotechnical work systems, integrating prevention into design projects for better adequacy with real-world operational requirements. Projects, provided throughout this book, take into account human work concretely, as well as health and safety aspects at work, using appropriate technology as support and service to safe, effective and comfortable human activity.

This book aims at promoting engineering work to the service of human work support, going from operations to design, considering all workers in the organizational chain or network (i.e., taking an experience-based approach). In addition, keys to a possible collaboration between researchers and practitioners should be provided, together with engineering designers and occupational risk prevention specialists, constantly keeping in mind that zero risk does not exist. Participatory design and co-construction of technology, organizations and jobs is at the heart of the contributions provided in the book, which presents ideas, experiences, stories, analyses and perspectives toward new possibilities of creating, deciding, designing and preventing risks.

The content of this book is rooted in two events where several experts, doers and thinkers, provided insightful information on risk-taking and management: a three-day conference organized by the European Air and Space Academy in February 2008 in Toulouse, France (Boy & Brachet, 2010); and a five-day workshop organized at Florida Institute of Technology in March 2014 (the next chapter of this book was written as a follow-up to this workshop, and introduces what we mean by risk-taking within an HCD approach). We extended this perspective to activity-centered work analysis, design and assessment within life-critical environments. More specifically, human activity will be considered from its dynamics and development aspects, especially in project management.

To summarize, Guy André Boy gives an introduction to the book (Chapter 1). Chapters 2–5 are about risk analysis and prevention design: the researchers' point of view for defining risk-taking contexts and a risk-prevention methodological framework useful for engineering design. Chapters 6–9 provide real-world experiences in safety-critical sociotechnical design: the meaning of risk-taking in practice. Chapters 10–13 present cross-fertilization of prevention and HSI for human-centered and organizational design: the orchestration of the integration of prevention and design. We tried to mix theory and practice, with fundamental models and real-life accounts. Our aim is epistemological, where we anchor models on concrete realities.

Grégory Michel (Chapter 2) provides a psychological and developmental approach in populations where risk-taking is incrementally learned. Risk-taking genesis starts when we are teenagers and continues during the rest of our lives. Not taking risk may be a source of risk because people will ignore what risk is about by a lack of situation awareness references, and sometimes lack of risk embodiment. How far can we go to find compromises between knowing more about risk in practice and not being exposed to irreversible situation where life is in danger?

Sylvie Leclercq (Chapter 3) proposes a model of risk-taking that enables to anticipate, manage and recover incidents and accidents at work. Risk-taking is about decision-making. We make decisions to find acceptable trade-offs among the possible actions to be taken in various industrial and medical environments. More specifically, grasping organizational complexity is one of the main difficulties in such decision-making activities. Residual risks that result from a lack of consideration in design are necessarily managed by people involved at operations time. The way risk is perceived is crucial. This is the reason why engineering design should anticipate risky situations to provide human operators with appropriate and useful affordances.

Neville Stanton and Catherine Harvey (Chapter 4) provide a critical view of human-error approaches. They propose a framework for systemic analysis of risk in teamwork. The resulting method called EAST (Event Analysis of Systemic Teamwork) enables to support analysis of social interactions in a team from a risk-taking perspective. In missile attacks, for example, it is crucial to identify and understand the broken links in the social network of the various human and machine agents. These broken links are sources of uncertainties and have direct repercussions on risk-taking of the various agents. EAST is case-based. It enables to support risk categorization and link appropriate strategies of actions to minimize danger.

Eric Drais (Chapter 5) reminds us that discussing risk-taking always involves an evaluation or judgement in a particular context. Faced with the daily need to design, adapt and control work organisation, risk management entails constant attention. To avoid a fast or simple attribution of risk factors, based on a case of road maintenance work, he proposes a methodology to support prevention work with a deeper analysis of individual and collective activity. Risk prevention work guarantees the development of a prevention culture and a renewed safety culture. Conversely, the Chernobyl and Challenger accidents in 1986 are examples of dramatic events caused by deviant groupthink and an insufficiently examined and discussed safety culture. The challenge in developing a more comprehensive risk prevention is to find a way to share knowledge and experience of risks within its contexts, in relation to safety culture. The implications for engineering design are crucial, where technology, organizations and jobs should be concurrently developed, using constantly updated experiential feedback.

Thus, design is seen as a never-ending process contributing of a living prevention culture.

Anne-Claire Macquet, Antoine Macquet and Liliane Pellegrin (Chapter 6) provide an analysis of high-performance sport activities for risk acceptance. They talk about the opportunity to win. Risk-taking is not only a matter of competence, knowledge and good practice where physical and mental preparation is key before the competition but also confidence and trust. Simulation is a major support for such preparation. Risk-taking should be learned while acting together with incremental rationalization of it – coaches play an important role there and debriefings are mandatory. Time pressure is one of the main drives in risk-taking at performance time; it should then be mastered.

Ludovic Loine (Chapter 7) provides his own experience, as a former submariner, on total isolation under water. What does it mean to take risk in such isolation conditions? Storytelling is a great way to identify categories of risks. How a safety culture in extreme conditions can be built, where the main issues are to live with people we do not know in a very closed environment? The concept of hierarchy and collaborative organizations is at stake, where safety experience is constantly capitalized and used. Team building is a major support for success of missions. Following procedures are constantly challenged with problem-solving activities, where crisis management is always at stake. People are always the last chance for avoiding, managing or recovering from accidents. This is the reason why competence and deep experience are assets for keeping safety a major principle. Such operational experience is a great background for HCD.

De Vere Michael Kiss (Chapter 8), a commercial airline captain and human-centered designer, promotes accident analysis as a contributor for HCD. Knowing about risk-taking in practice is an incomparable asset in HCD. More specifically, knowledge about concrete pilots' capabilities and limitations is essential in HCD not only to better identify and formalize assessment metrics for formative evaluations in human-in-the-loop simulations but also to propose engineering design alternatives. For example, prescribed team-oriented leadership style is a practice of the US airline industry, which includes charisma, inspiration, mental stimulation and consideration of individuals. This style is an important behavioral good-practice, but it is crucial to know and understand the difference between such task-based prescription and the real activity of pilots. This is precisely such knowledge about pilot's activity and resulting team coordination that should be considered in HCD.

Anabela Simões (Chapter 9), a specialist on aging at work, presents a life-long approach of risk-taking in life-critical environments. Human operator's aging is influenced by both internal and external factors. Incremental training is key in learning what risk-taking is about (i.e., we need to learn new practices at various moments in life). Human capabilities change during our life cycle. Some of them must be discovered because they are specific to individuals, other can be anticipated because they tend to be generic across the population. Some factors, such as fatigue, capability to handle workload and physical performance, are changing when aging. Consequently, performance is always adapted to these capabilities (e.g., situation awareness, decision-making and action-taking). Therefore, people do not take risk in the same way with respect to these capabilities. It takes time to adapt expertise with respect to aging, but when mastered, new expertise enables to handle risky situations anyway.

Chapter 10 was written by R.B. Wears at the very beginning of the project, but it could not be completed and revisited because the author passed away. Robert Wears was a physician and critical care specialist, who claimed that risk-seeking in health care is a life-critical necessity. Given the importance of his work and purpose for this book, we wished to keep his text as he written. Also, a French colleague who is a specialist in hospital work had proposed supporting and completing his work after this archive.

In his chapter, Robert separates health and safety but considers the trade-off between short-term and long-term risk management. He discusses regarding emergency care, for example, three important requirements: airway problem management and how life maintenance can be taught in emergency situations (i.e., the physician needs to make sure that the patient keep breathing!); violating procedures when they are not working in context and making sure that the group knows and accept the resulting practice (i.e., knowing when the physician should shift from procedure following to problem-solving); and in desperate circumstances, such as when airways are blocked, radical solutions, such as intubation, should be implemented. Coordination among services should be thought in anticipation, instead of requiring heroic acts at the last minute in emergency situations. Consequently, flexibility in such situations should be better anticipated at design time, from three concurrent perspectives: technology, organizations and people's competences that we need to support and give means to be developed.

Lucie Cuvelier (Chapter 11) develops a deeper follow-up of Robert Wears' introductory account. Anesthesia is a life-critical process that involves advanced technology, organization and competence. What makes sense here is risk-seeking and the paradox of variability in health care. The goal of anesthetists is keeping patients breathing in good conditions, and more fundamentally keeping them alive. Variability can be expressed in terms of patient conditions (e.g., degree of injury and disease) and external conditions (e.g., time pressure and organizational constraints, such as the number of critical care patients at a given time). Again, preparation and experience are key factors for success in emergency situations. Activity-centered design must be further developed as a participatory process involving human factors specialists, medical personnel and engineering designers.

Sébastien Boulnois (Chapter 12), a human-centered designer, with the help of Edwige Quillerou, an occupational psychologist, presents risk-taking in the context of a design project of a navigation onboard system, called OWSAS (Onboard Weather Situation Awareness System). This chapter presents the education and training of a human-centered designer, who worked with a work psychologist and pilots. All these competencies were incrementally integrated to produce a system that considers activity-based requirements. Five key moments (KMs) were considered and implemented: (KM1) interviewing pilots on flight conditions; (KM2) confronting experience and knowledge of two pilots; (KM3) offline prototype assessment by pilots; (KM4) organizational context analysis; and (KM5) performing early-stage using human-in-the-loop simulation, therefore considering human activity. All KMs are built with pilots.

Leïla Boudra, Pascal Béguin and Valérie Pueyo (Chapter 13) provide an approach for the design of "plastic" work systems in the waste management and recycling sector. The connotation of plastic work system is aligned with flexibility that was already introduced elsewhere (FlexTech by Boy, 2020). What is addressed here is both connected work-related prevention and sustainable work systems. Coordination should be developed to concurrently handle political decisions and technical capabilities. Decision-making is strongly influenced by both systematic work analysis and discussions among waste management workers. The production model should then be adapted accordingly. Plasticity is improved by a better integration of design and prevention approaches and actual work to the benefit of concomitant consideration of health and safety, and more fundamentally improving HCD.

Summarizing, in initial chapters (2-5) of the book that there are many benefits to consider regarding prevention in engineering design at all stages of the life cycle of a sociotechnical system. Risks should be conceptualized, analyzed and understood in various purposeful ways. In chapters 6-9, we see that expert knowledge and experience are crucial in risk prevention for engineering design by considering various kinds of activity natures and analyses. Finally, chapters 10-13 provide a synthesis that discusses the integration of various kinds of specialists (e.g., expert analysts, subject-matter experts and prevention specialists) for participatory design.

The conclusion is a departure toward integration of several isolated practices and theories. First, research and practice should be better integrated, leading to experience-based associated with creativity. This requires taking an epistemological approach that enables us to better understand the genesis of solutions, which are at the crossroad of technology, organizations and people's jobs co-design. Second, being clear on what purpose and model of human activity we are referring to is crucial. Third, developing a systemic approach in the perspective of transforming work is crucial toward improving the quality of life.

Edwige Quillerou and Guy André Boy

About the Editors

Guy André Boy, Ph.D., is a cognitive scientist and engineer, fellow of INCOSE (International Council on Systems Engineering) and the Air and Space Academy. He holds the FlexTech chair and is a university professor at CentraleSupélec (Paris Saclay University) and ESTIA Institute of Technology. He was chief scientist for human-centered design at NASA Kennedy Space Center, university professor and dean at Florida Institute of Technology, where he created the Human-Centered Design Institute. He was senior research scientist at Florida Institute for Human and Machine Cognition and former president and CEO of the European Institute of Cognitive Sciences and Engineering (EURISCO), France. He is the chair of the Aerospace Technical Committee of IEA (International Ergonomics Association) and INCOSE Human Systems Integration Working Group worldwide.

Edwige Quillerou, Ph.D., is an occupational psychologist and health & safety scientist at INRS, the French National Research and Safety Institute for Prevention of Occupational Accidents and Diseases. She contributes to both research and field interventions in a variety of design and work organization projects. She is involved in interdisciplinary collaborations with researchers in the human and social sciences as well as engineers with a view to improving the organization of prevention with respect to occupational risks through improved engineering design. She is a member of IEA (International Ergonomics Association) and INCOSE (International Council on Systems Engineering).

Contributors

Pascal Béguin, Ph.D., is a full professor at the Institute for Work Studies of Lyon, previously director of Research at the French National Research Institute in Agriculture, and he has been invited professor in many universities (mainly to Helsinki, Finland; Hobart, Australia; and Rio de Janeiro, Brazil). He is also the founder of the open access journal Activités (https://journals.openedition.org/activites/). His work is aimed at contributing to design and change at work. He has published, among others, a book *Risky Work Environments: Reappraising Human Work within Fallible Systems* with C. Owen and G. Wackers (Ashgate, 2009).

Leïla Boudra, Ph.D., is a researcher and lecturer in Ergonomics, currently postdoctoral researcher at the CNAM-CRTD, Paris. She also coordinates a network of researchers on design in ergonomics for sustainable development. Her work targets the design of sustainable work systems, questioning the development processes of individuals, groups, and work-related prevention.

Sébastien Boulnois, Ph.D., is currently a researcher at Round Feather, Inc. in San Diego, California. He is specialized in applied qualitative research, both generative and evaluative, in many fields such as health care and tech. He owns a Ph.D. in Human-Centered Design from Florida Institute of Technology in Melbourne, Florida. Before he decided to dedicate his work life to study the Human, Sebastien was focusing on Systems Engineering with a background in Engineering Science and Human-Machine Systems.

Lucie Cuvelier, Ph.D., is a Health and Safety engineer. She is an assistant professor in ergonomics and psychology at the University of Paris 8, France. Her research aims to develop new models to improve systems safety, with emphasis on competences development and knowledge construction (e.g. full-scale anesthesia simulator for training). She is interested in a variety of work dimensions in the fields of occupational health, system safety, and organizational reliability.

Eric Drais, Ph.D., sociologist, is a senior research scientist at INRS, specialized on health and safety management at the workplace and working life conditions organization in companies. He was part-time University professor for 15 years at Paris Saclay University and managed comparative studies on health and safety management systems and on safety regulation toward nanomaterials risks. His work focuses specifically on how organizations shape both the social production of risk and health and safety cultures. Eric Drais participated in several national or international projects and research efforts: OSHA Europa, ANR Nanonorma, and ISO 45001.

Catherine Harvey, EngD, is an assistant professor in the Human Factors Research Group at the University of Nottingham. She held an Anne McLaren Research Fellowship between 2013 and 2020 and moved into an academic role in 2020. Catherine's academic background is in usability evaluation, human factors methods and HCI and she has applied this knowledge in a variety of domains, including transport, sociotechnical systems, and aviation. She has an Engineering Doctorate (EngD) from the University of Southampton (2012) and a BSc in Ergonomics/Human Factors (Loughborough University, 2007).

De Vere Michael Kiss, ATP, Ph.D., is an airline captain/instructor, scientist and specialist in: human factors engineering, crew resource management, leadership and human centered design. He is teaching at the Florida Institute of Technology and a member of the American Institute of Aeronautics and Astronautics, the American Astronautical Society, the American Psychological Association, the Association for Aviation Psychology, and the Society of Aviation and Flight Educators. He is also the chairman of the Society of Automotive Engineers G-10G committee making vital recommendations to the U.S. Federal Aviation Administration regarding Airline Pilot Training.

Sylvie Leclercq, Ph.D., is a researcher at INRS, the French National Research and Safety Institute, whose missions are the prevention of occupational accidents and diseases. Her research focuses on the prevention of occupational accidents, in particular on the risk of occupational accidental injury if movement of the worker is disturbed when carrying his/her task.

Ludovic Loine, is a nuclear engineer trained in the French Navy Schools. After 16 years as submariner, he joined engineering field to implement Human-Centered Design concept and managed Human Factors Engineering (HFE) for several clients during numerous years. He is at present CEO of SAS AXION. As expert consultant in HFE and project manager, he is leading HFE on projects in the nuclear and oil and gas industries.

Anne-Claire Macquet, Ph.D., HDR, is a senior lecturer in both Cognitive Ergonomics and Human Factors, and Sport Psychology. She currently works in the laboratory of Sport, Expertise and Performance EA 7370 within the French Institute of Sports. Her work includes decision-making in dynamic and complex environments including those with high stakes and risks relating to top level competitions such as World Championships. She also undertakes consultancy activities in coaching and ergonomics as a pathway to the individual and collective performance and well-being of elite athletes, national coaches and staffs (https://orcid.org/0000-0003-2554-0540).

Antoine Macquet is an urban planner specialized in risk and crisis management. He is currently working as a flood specialist for the *Etablissement Public Seine Grands Lacs*, the local authority in charge of flood risk for the Paris metropolitan region, France. His job is to advise authorities of towns and cities on emergency and contingency plans regarding natural disasters. He is also a lecturer for Paris 1 Panthéon Sorbonne University Master in Risk and Emergency Management.

Grégory Michel is a professor of Psychopathology and the head of Psycho Criminology Axis of Institute of Criminological Sciences and Justice in the University of Bordeaux, France. He is an expert in risk taking behavior and antisocial conduct among young people and early psychological intervention. He is also a clinical psychotherapist and psychologist expert at the French court. He had been the head of the Psychology, Health and Quality of Life Laboratory in Bordeaux, and director of the Master of Psychopathology. Currently he is an associate researcher at the International Laboratory GRIP (Group for Research and Psychosocial Intervention) of Universities of Montreal. Gregory Michel is a scientific expert for risky behaviors and violence with the Educational Ministry as well as for the Social Sciences and Research Council of Canada. He has published more than 95 articles in indexed journals, and five books regarding conduct disorders and risky behaviors. He is a founding member of the Mental Health and Prevention Editorial Board. He has trained numerous students and fellows and has obtained numerous grants for research projects.

Liliane Pellegrin, Ph.D., is a senior researcher in cognitive ergonomics, human factors linked to an expertise in public health and epidemiology. She is a member of the Centre d'Epidémiologie et de Santé Publique des Armées (CESPA) of the Service de Santé des Armées (SSA), within the MDBS department (Modelling, Data Science, Biostatistics, and Information Systems). She oversees a simulation unit, developing training in public health surveillance, and epidemic management. She works on individual and collective expert decision-making in public health, more particularly on epidemic surveillance activities in current and crisis situations (https://orcid.org/0000-0001-8947-7056).

Valérie Pueyo is a professor at the Institute for Work Studies of Lyon. She is an active member of scientific councils (for #DigitAg a Digital Agriculture Convergence Lab, and INRAe – Division Sciences for action and transition). She is also vice president in charge of International Affairs for the French Speaking Ergonomics Society. Her work is aimed at contributing to design and change at work in the perspective of the Anthropocene. She is co-author among others, of a book *Concevoir le travail: le défi de l'Ergonomie* (Octarès, 2021).

Anabela Simões obtained a Ph.D. in Ergonomics at the former Technical University of Lisbon, which is now the University of Lisbon. She is a full professor at the LUSOFONA University of Lisbon and has participated in several research projects funded by the European Union since 1992, whose central theme was aging. She was president of the Portuguese Society of Ergonomics (APERGO) from 1997 to 2003 and a member of the IEA board since 2000. She is also a member of HUMANIST (Human-Centered Design for Information Society Technologies), a European Research Network acting as a virtual center of excellence.

Neville A. Stanton, Ph.D., D.Sc., is a chartered psychologist, chartered ergonomistc, and chartered engineer. He is a professor Emeritus in human factors engineering at the University of Southampton in the United Kingdom. His research interests include modelling, predicting, analyzing, and evaluating human performance in systems as well as designing the interfaces and interaction between humans and technology. The University of Southampton has awarded him a Doctor of Science in 2014 for his sustained contribution to the development and validation of human factors methods.

Robert L. Wears, professor in emergency medicine at the University of Florida, USA, was a senior physician and leading international expert in patient safety. He was a visiting professor at the Imperial College, United Kingdom, senior associate dean for hospital affairs and former chair of emergency medicine. He was committed to applying state of the art theory and technologies to improve patient safety. He died in 2016 at the beginning of this book project. We would like to pay a special tribute to his contribution for the hospital's world. Our civilisation has never been more aware of the importance of taking care of its hospitals. We are grateful to him for having worked for this.

Chapter 1

Risk-taking: A life-critical necessity

Guy André Boy

Paris Saclay University (Centralesupélec) and
ESTIA Institute of Technology
Gif-sur-Yvette & Bidart, France

Are there actions that are not risky? As soon as we are born, even if we don't realize it, we take a risk. Are we going to survive? Progress in medicine contributed to increasing the probability of survival at any stage of our life. Of course, this probability depends on where you were born on the planet Earth. When the baby starts experiencing walking, he/she takes risks and finally discovers balance on his/her two feet. This process of trial and error contributes to learning. It could be supervised or not, but it seems that it came naturally for most of us. Later, during our lives, we take risks for several purposes, whether it is for learning, doing our job, helping other people, or just for fun.

There are professions that involve risk-taking more than others. Astronauts, for example, are selected to be able to survive in environments that are not usual, and where unexpected events lead to difficult problem-solving activities. This kind of job is commonly supported by procedures, but uncertainty and complexity of the environment require operational skills that are not present in everyday activities on Earth. Another example can be found in hospital operating rooms where surgeons need to use their experience and sometimes invent new protocols to save lives. As space, medicine can be strongly based on procedures but require people who are trained to handle unexpected events. On another register, nuclear power plants are highly automated systems controlled by human operators who also use many procedures. In all these domains, there are circumstances that require decisions to be made that involve trading among various kinds of risks, especially during a major crisis.

These three examples show that risk is related to various concepts ranging from **procedure following** when situations are recognizable and well understood to unconventional decision-making and creativity when we must **deal with the unexpected**. This book presents several approaches and methods that can be useful to and usable by risk takers and everybody who needs to design technology, organizations, and jobs in risky environments. It also tries to open a new field of research, where risk has not only to be prevented and managed but also to be taken. Dealing with life-critical systems requires careful preparation, as well as availability and courage to

DOI: 10.1201/9781003221609-1

act. No matter the precautions that you take in dangerous situations, the "Risk zero" never exists! When the likelihood of successful action is not well known, you will need to have a set of well-identified recovery spaces where you can "land" in case of failure. Risk-taking belongs to the art of abduction, which is the art of guessing a possible outcome and verifying it by acting appropriately. Risk-taking deals with domain expertise and embodied skills that enable you to have the right heuristics at the right time to abduct the right outcome.

Three cases of risk-taking could be distinguished (Boy, 1993). Imagine a system that suddenly fails. First, if the system works in open loop and is interruptible, it can be repaired in a static mode. Risk is minimal in this case. This could be the case of a coffee maker that suddenly does not work and can be repaired by turning the power off, for example. Second, if the system works in close loop and is interruptible, it can be stopped in some conditions. For example, a car that has a sudden engine failure and rolling in a very busy street. Conditions to stop the car can be finding an appropriate spot and constantly monitoring pedestrians before stopping. Once the car is parked correctly, repair can start safely. Third, if the system works in close loop and is uninterruptible, contribution to repairing or at least coping with the failure should be done without stopping the system. This is the case of airplanes.

For example, landing on the Hudson River is an option that Captain Sullenberger took after a flock of Canada geese stroke the aircraft that was immediately disabled, while taking off from LaGuardia Airport on January 15, 2009. Making the right decision, at the right time, is always key in such situation. This requires three main ingredients: knowledge and skills, hard work, and luck. That day, Captain Sullenberger had the three "green"! He is a competent, very skilled pilot, and more importantly very aware of safety issues. He was totally in the control loop using his excellent gliding skills. Luck was there too, since there were no big waves on the Hudson River, and landing ("rivering") was straightforward for him. Taking the risk of landing on the Hudson River instead of returning to LaGuardia was a big decision. He was aware of the various possible outcomes, and his abduction turned out to be the right one.

The concept of risk needs to be further defined. The Oxford Dictionary for English usage defines "risk" as a situation involving **exposure to danger**. Risk is associated to responsibility and insurance. Anytime we take a risk, we are responsible (i.e., accountable, liable) for the outcome in the legal sense. Therefore, exposure to dangerous situations should always be assorted to various kinds of insurances. The first one consists in skills, competences, and knowledge about the risk to be taken. It is usually made concrete in the form of redundancy (i.e., passive stability that enables safe return to a stable state) and abilities (i.e., active stability that constantly keeps a stable state). Other types of insurances could also be taken regarding the consequences of accidents; we are in the classical business of insurance companies. In any case, taking a risk proceeds in the frame of mind that there is a chance of something unpleasant may occur... as a matter of fact pleasant also in some cases! In the former cases, you can be dead; in the latter, you can be a hero!

Risk should be defined in terms of management of the variability of the course of events around us. Sometimes these events are unexpected. Unexpected events can be known in advance, and in this case, the use of procedures and situation awareness is a major asset to deal with them, or not known at all, and in this case, creativity is at stake. **Creativity is a matter of synthesis and integration** of things that are already known. Creativity should provide meaning. Sometimes it takes try-and-error activity to figure out what is going on and what could and should be done. This takes talent, competence, and chance.

Good risk takers have necessarily taste for exploration. They tend to increase their performance when the perception of risk increases (i.e., this is the opposite for regular and risk takers averse). In other words, we cannot really know a risk if we did not see its triggering point and outcome close enough. Risk takers know their limits in terms of possible recovery strategies in case of failure. They build **intrinsic redundancies** (i.e., they learn and train to build useful and usable skills and competencies) and **extrinsic redundancies** (i.e., they learn about their environment to know about possible recovery spaces). Risk-taking can be individual as we already talked about, but it can also be collective. In this case, taking a risk is a team effort that needs to be prepared. One of the major factors is building confidence within the team. For example, each team member should be able to be a leader and a follower when necessary. Finally, there cannot be risk-taking without enthusiasm. In particular, the team should be able to share the optimism of the action and constant verification that the abduction process is appropriate.

As a logical inference, **abduction** assumes that an outcome (a consequence) is true and leads to the demonstration that the inference from the current state to this outcome is also true. For example, John F. Kennedy claimed in 1961 that Americans will walk on the Moon before the end of the sixties and come back to Earth safe. NASA managed to make true the underlying inference. Risk-taking as abduction is then related to intuition, creativity, courage, and boldness. In addition, we cannot take risks without openness, humility, initiative, and non-conformism.

"The cold speculative intelligence comes to confirm the brilliant intuition of the artist... There is something artistic in scientific discovery and there is something scientific in that which the naïve call 'brilliant intuitions of the artist'. What they share is the felicity of Abduction" (Eco, 1990). We can see abduction as a form of intuitive reasoning that consists in removing improbable solutions. It is typically supported by intuition, pragmatism, experience, and competence.

Risk-taking can be both goal-driven and event-driven. When it is goal-driven, risk takers must manage to make the decided abduction work.

The problem with this strategy is that there is no guaranty of success. We need to believe in the vision. We need competence, coordination, enthusiasm, and dedication. When it is event-driven, risk takers must constantly predict what could happen next. The problem with this strategy is that prediction is always short-term and, therefore, may introduce loss of track by optimizing locally and losing the holistic endeavor. We can say that the decision to go to the Moon was goal-driven, and Americans managed to do it right and succeeded. We also can say that it was event-driven because this was an answer to the Soviet Union that was ahead in terms of Space exploration at that time. Human being always swaps from goal-driven to event-driven behaviors and strategies. This capacity is often called opportunism. Nevertheless, when the center of gravity of our behavior is too much on one side, goal-driven or event-driven, we tend to enter and stay in a neurotic circle. Today, for example, some countries are so frightened to take any risk that they do not commit to any long-term vision. They fall into such a neurotic circle of "short-termism," trying to predict the future, which is unfortunately impossible except in the very short term. They are only event driven.

Could we really learn when we are forbidden to make errors? No! During the last three decades or so, we tried to become error prone. All this was done to avoid risk! Human errors were the main concern for human factors people. We did everything possible to avoid human errors or diminish their consequences. What did we learn? We learned how to adapt machines to eradicate human errors, committed before this technological adaptation. However, since *"errare humanum est"* is always true, people now commit new types of errors using these new systems. We automated one layer after another to create systems that were difficult to understand because of the accumulation of software layers. The *leitmotif* was and still is "reducing risk" to the point of "do not take risk any longer." This risk aversion led to taking even bigger risks by avoiding considering complexity as it really is. In this book, we will analyze the difficult problem of stall in flight. This is a good example of a nonlinear phenomenon that cannot be oversimplified. Understanding its complexity is mandatory for any pilot, to the point that the pilot needs to become familiar with it. How? There are not many possibilities. One possibility is "understanding it theoretically" – for this, a minimum of mathematics and physics are necessary. Another one is feeling it at least a few times to embody the phenomenon. Are we prepared to do this? It does not seem that we are any longer! We need to remember the success of the Apollo program more than 40 years ago, which started with a dramatic loss of Apollo-1 crew, burned on the launch pad in their capsule. This event did not stop the program that led humanity to the Moon. Today, we do not want to take any risk. In addition, school does not prepare us for such meaningful theoretical developments and for complexity science. At school, we learn linear algebra and simplification of the world instead of attacking nonlinear systems as they are.

Adaptation is a crucial issue in risk-taking. It usually takes a fair amount of time for people to adapt to complex safety-critical systems. Once they are fully adapted, they usually do not want to change. They become conservative. Today, we are at the crossroad of a new renaissance, where new kinds of systems are emerging from the massive amount of technology that we constantly integrate. Our world changes very rapidly. Consequently, conservatism is no longer possible. This is the reason why new risks are taken, and we need to experience them to figure out what is really going on. This does not mean that we should avoid precautions, especially when we have poor knowledge of the consequences of some of our actions in a constantly changing environment. We just need to experience it, explore it, understand it, and rationalize it. Instead of a precautionary principle, I prefer a principle based on knowledge, experience, and action. We need to investigate and become familiar with the complexity of our new interconnected information-intensive world, instead of to be frightened by its complexity and the possible danger it may induce.

Risk-taking is about projecting yourself into the future. I cannot resist quoting Alan Kay at this point, "the best way to predict the future is to invent it." As already said, prediction is only short term to make sense. Risk-taking needs to deal with ignorance and uncertainty, and certainly not with probability when the expected outcome is expected to be disastrous. It is interesting to observe that risk is often predicted using the following formula:

Risk = (probability of an accident occurring)
    × (expected loss in case of accident).

The main issue with this formula comes when the probability of an accident occurring is very small (i.e., tends to zero, mathematically speaking), and the expected loss in case of accident is very big (i.e., mathematically speaking tends to infinity). In mathematics, zero multiplied by infinity is undetermined. On this example, we see that the problem comes from the theory itself, where the concept of probability is not appropriate. Instead, the concept of possibility and necessity would be much better because it introduces a confidence interval (necessity, possibility) that determines the **ignorance** on the phenomenon being studied. Risk takers should reduce this ignorance/confidence interval. If we take the Fukushima accident, for example, there were very few earthquakes followed by tsunamis of the same amplitude in the past, making the probability of an accident occurring very small, but the expected loss in case of accident very big. Instead, the possibility of an accident occurring in the Fukushima area is close to 1 and its necessity very small. If we calculate the risk as a product of the possibility by the expected loss, it is the expected loss in case of accident that counts. This means that building a nuclear power plant there was at big risk.

This brings to the front risk-taking in **crisis management**. A crisis is defined by its rise, a break (usually a surprise), and a transformation of the overall system, organization, or environment. Crisis management must deal with a threat that is usually not well anticipated. There are very few standard operating procedures. Problem-solving is the best resource. Consequently, crisis managers need to be well informed of what happened, what is going on, and what could happen soon. We could call this state awareness, where people involved not only require knowing about the situation but also about its cause and constant evolution (i.e., various levels of state derivatives in the mathematical sense). A recent research effort showed that 3D visualization mixing moving images of a crisis in real world with digital artifacts (e.g., diagrams, purposeful computer aided design [CAD] plant representations, rescue reports) is a good solution supporting crisis management. Indeed, crisis managers must take risks when there is no concrete procedure to help them to act. Therefore, they need a good mix of field reality and integrated parameters and technical models.

Since risk-taking deals with complexity, we need to better define what a complex system is. A complex system can be defined as a "**system of systems**" that has two main characteristics: emergence over scale and self-organization over time. In addition, we cannot define complex systems in only one way; we need to be informed using several perspectives, which include systems theory, nonlinear dynamics, pattern formation, evolution and adaptation, networks, collective behavior, and game theory. Let's try to give some insights on each of these perspectives.

Systems theory informs us on the fundamental phenomenon of **homeostasis**, initially discovered by Claude Bernard, a French physiologist (1813–1878), who coined the term preservation of internal environment (*milieu intérieur* in French) to make explicit human body regulatory processes such as healing. Walter Cannon, an American physiologist (1871–1945), who took over Bernard's work, coined the actual term "homeostasis" in 1926. Later, Norbert Wiener, an American mathematician and philosopher (1894–1964), was inspired by this initial grounding work and invented cybernetics, which installed a formalization of the concept of feedback and system dynamics. Cybernetics had a strong influence on automatic control, computer science, biology, philosophy, and society. The notion of complex system started to evolve toward self-regulation (i.e., a system that constantly uses the feedback of its output to keep an assigned set point). This is the essence of homeostasis. A step further, Humberto Maturana, a Chilean biologist and philosopher (1928–present), developed with his student Francisco Varela, a biologist philosopher and neuroscientist (1946–2001), the concept of **autopoiesis**, which is an extension of the concept of homeostasis. Autopoiesis refers to systems that can keep their internal environment and reproducing themselves. It is applied in biology, chemistry, systems theory, and sociology.

Why are systems theories are important in risk-taking? Basically, risk-taking is intimately related to the notion of affordance, which is the relation between the risk taker and his or her environment that affords the opportunity for the risk taker to perform a risky action. For example, a clear sunny day affords the skilled sailor to navigate full speed on the ocean. When the sunny day transforms into a nasty storm, this kind of affordance becomes more critical because appropriate actions become mandatory. An affordance is not a property of either the risky environment or the risk taker (e.g., ocean or skilled sailor), but a relation between the two. Affordances are intimately related to *enaction*, which is the embodied interaction between people and the physical world, as Francisco Varela, Evan Thompson, and Eleanor Rosch defined it (Protevi, 2006). For that matter, systems theories that include homeostasis, cybernetics, and autopoiesis are important to understand and master for any risk taker.

Risk takers need to know the nature of **nonlinear dynamic systems** to better anticipate their behavior and act on them. For example, René Thom, a famous French mathematician, invented and developed the catastrophe theory (Thom, 1983), which is more a language that enables to organize experimental data in various appropriate conditions, like Newton's method of fluxions that initiated differential calculus (Kitcher, 1973). Differential calculus was created to describe the evolution of system states, and particularly the evolution of the movement of a body. Thom's catastrophe theory attempts to describe discontinuities of such evolution. In fact, the main idea is to jump from a continuous evolution in a given space to another continuous evolution in a different space. A typical example is the transition from a particular phase to another phase, e.g., the transition from gas to liquid or from liquid to solid. This phase change, or more generally this space change, brings us from one evolution to another involving different attributes and properties.

In linear systems, we always expect that a problem led to a single solution. In the real world, solving a problem is analyzing **several possible solutions** and selecting the best one with respect to principles and criteria formulated as well as possible. Risk-taking in nonlinear dynamic systems or environments requires appropriate **heuristics** that are only learned through experience and very specific advanced human capabilities. There is always a tradeoff between using very stable heuristics, keeping a conservative attitude, and projecting ourselves into the future and managing the situation until goals are realized. In both cases, we take risks. The conservative attitude may block our situation awareness of changes and/or avoid us to take any untested initiatives. The risk in this case is lack of adaptation to our changing world. Conversely, the visionary attitude forces us to be extremely knowledgeable, hyper-available, and very well trained to solve problems as they come. In other words, our **problem-solving** skills should be at their best. The former is procedure based, and the latter is problem-solving based.

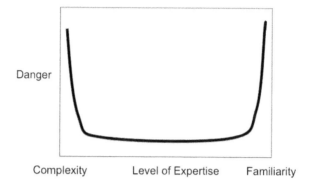

Figure 1.1 Danger versus level of expertise.

Risk-taking happens when the situation is **complex, in the sense of unfamiliar.** Therefore, we always analyze an unfamiliar situation before doing anything. Once it becomes familiar, risk is considerably decreased. However, this may not hold forever because when we are too familiar with a situation, activity becomes routine, and we tend to be complacent (Figure 1.1).

We can become very vulnerable in routine situations because vigilance may decrease considerably as over-trust increases. High level of expertise does not deny the fact that we need to remain vigilant. Consequently, taking a risk requires availability, attention, and awareness always. This requirement tends to increase workload and time pressure. Workload remains a very difficult factor to study because it is both an output variable and an input variable. In other words, we produce workload, and we need workload to have a reasonable level of arousal to act appropriately. When workload is too high, we are likely to be stressed; when workload is too low, we are likely to be hippo vigilant. Therefore, it is crucial that people keep a "reasonable" level of workload, between stress and hippo-vigilance, to be effective in their job. This is even more crucial for risk takers.

We want to deconstruct complexity using modeling and simulation (i.e., we want to see what was supposed to be complex in a different "simpler" way). For example, a long time ago people used to watch the sky as a complex set of stars; Copernic, Kepler, and Galilée deconstructed such complexity by providing models based on two simple variables, i.e., distance and mass. We are looking for models that describe what scientists observe (i.e., they rationalize complex datasets). This kind of rationalization usually consists in finding persistent behavioral patterns, identifying generic phenomena, and constructing relationships among components and so on.

We will then develop the following points (each being a potential chapter):

- situation awareness and decision-making;
- courage, boldness, and action;

- creativity as integration and synthesis;
- procedure following and problem-solving;
- training for risk takers;
- effect of aging on risk-taking;
- risk-taking in emergency medicine;
- risk-taking in nuclear energy management;
- risk-taking in nuclear power plants and process control;
- risk-taking in aviation;
- risk-taking in space and extreme environments;
- risk-taking in automobile;
- risk-taking in crisis and disaster management;
- abduction as a central cognitive process for risk-taking

This book was triggered by many discussions that emerged from two former events on risk-taking. The first one was a conference organized by the Air and Space Academy in Toulouse, France, on February 4–6, 2008. A synthesis booklet was produced (Boy & Brachet, 2010). The second one was a workshop on risk management in life-critical systems, organized in Melbourne, Florida, on March 3–7, 2014, within the scope of the Partner University Funds (PUF) partnership program between Florida Institute of Technology and University of Valenciennes. Some of the chapters' authors participated in these events.

REFERENCES

Boy, G.A. (1993). Knowledge acquisition in dynamic systems: How can logicism and situatedness go together? In N. Assenac, G.A. Boy, B. Gaines, J.G. Ganascia, Y. Kodratoff & M. Linster (Eds.), *Proceedings of the 7th European Knowledge Acquisition for Knowledge-Based Systems Workshop (EKAW'93)*, Lecture Notes in AI, Springer Verlag, Berlin, ISBN: 3-540-57253-8.

Boy, G.A. & Brachet, G. (2010). *Risk Taking*. Dossier of the Air and Space Academy, Toulouse, France. ISBN 2-913331-47-5.

Eco, U. (1990). *The Limits of Interpretation (Advances in Semiotics)*. Indiana University Press, Bloomington.

Kitcher, P. (1973). Fluxions, Limits, and Infinite Littleness. A Study of Newton's Presentation of the Calculus. *Isis* 64 (1), pp. 33–49. doi 10.1086/351042.

Protevi, J. (Ed.) (2006). "Enaction". *A Dictionary of Continental Philosophy*. Yale University Press, London, UK. pp. 169–170. ISBN 9780300116052.

Thom, R. (1983). *Parabole et Catastrophes*. Flammarion, Paris. ISBN 2-08-081186-X.

Chapter 2

From praise of danger to "reasoned" risk-taking

A psychological approach to risk-taking

Grégory Michel
University of Bordeaux
Bordeaux, France

CONTENTS

Risky behaviors that mark every stage of human life occur either intentionally and are thus considered as risk-taking (e.g., making risky decisions despite the inherent danger, or the deliberate quest to find oneself in a worrying situation), or unintentionally and are then taken to demonstrate imprudence or negligence (e.g., errors, disempowerment, taking poor decisions, lack of control, not respecting the rules, etc.). Whether it is risk-taking or recklessness, dangerous behavior combining both human and organizational factors may have many consequences on one's well-being, health, and success. Therefore, given the consequences in the short, middle and long term (diseases, accidents, social disruption, repercussions at work, death, etc.), it is important to consider the issues or the determinants of these risky behaviors, especially in sports, in the area of accidents and even in the field of civil safety. This chapter adopts a psychological approach to the relevance of the concept of risk-taking, highlighting the human and health factors involved. In particular, the concept of "rational" risk-taking is developed.

2.1 WHAT IS A RISK?

The etymology of the word "risk" refers to two concepts. The first comes from the Latin verb *resecare* meaning "to cut in two", "to separate". Risk is what cuts the individual in the event of failure or error. However, risk-taking is also cutting oneself off from the known, the safe environment in which we live, by confronting an unknown universe. The second etymologic origin extends this existential concept since it comes from the Greek *rhizhikhon*, which itself comes from *rhiza* meaning "root". This relates risk to the original and to some extent to oneself. By exposing oneself to risk, the subject calls upon his own resources, his physical and psychic abilities to confront danger. Risk-taking in terms of these two etymologies refers to the need to cut free from a reassuring protective environment in order deliberately to confront a world that is not reassuring with a view to testing one's physical (e.g., in the field of risky sports), emotional (e.g. taking the risk of being in a distressing situation) or intellectual (e.g. taking the risk of thinking differently from the group) capacities.

Most authors agree that a risk is a danger whose outcome can be anticipated up to a point and which one can more or less predict (Kouabenan et al., 2006; Michael, 2001). Taking a risk means venturing out, taking a chance. The risk is related both to the dangerousness of the activity and to the negative consequences that the individual may suffer (loss of money, having an accident etc.). Risk is an entity that concerns both: (i) "stochastic" random processes depending on the situation (e.g., human, mechanical failures, weather conditions, etc.), and (ii) situations comprising a challenge between gravity of the situation and negative consequences (e.g., risk of accident, loss of money, lethal risk, etc.). Some authors emphasize the social nature of risk behavior. For example, Turner et al (2004) define such behaviors as resulting either from a socially unacceptable volitional behavior with negative consequences in terms of morbidity or mortality when precautions are not taken (such as, abuse of toxic substances, excessive speeding, drinking, and driving) or a socially accepted behavior in which the danger is recognized and valued (competitive sports, parachuting, etc.).

Furthermore, it is customary to distinguish between short-term risks (immediate mortal risk) and long-term risks (deferred mortal risk). In the case of short-term risk activities, behaviors involve the concept of an act more than in the second type of risk. These acts can be considered as falling within the somatomotor register: sports involving physical risk, risk-taking in a motor vehicle, etc. In this case, there is a "risk penalty" with little chance of reversal: either one experiences victory or it is failure in terms of accident or death. However, the long-term risk or delayed risk reflects the potential danger that is usually inherent in a repeated activity. This type of risk mainly concerns the consumption of psychoactive substances. Indeed, in the consumption of toxic substances, the risk is above all perceived and defined in terms of the mental and physical mechanisms of addiction.

Nevertheless, the consumption of psychoactive substances does not only involve delayed risk because substance abuse and its harmful consequences (violence, but accidents also) and the risk of overdose in drug addiction fall within the short term. Similarly, with behavior that carries an immediate mortal risk, some individuals may potentially develop a relationship of dependence on it. Indeed, in previous work, we showed that some athletes who undertake extreme activities practice them even more and push back the safety threshold even further by adopting increasingly risky behavior to experience even more intense emotions, the danger being that they may develop an "addiction to danger" (Michel et al., 2010).[1]

2.2 HEURISTIC VALUE OF THE NOTION OF RISK-TAKING

Any conduct involves risk. Since zero risk does not exist in terms of probability, one can talk more of tolerated risk (i.e., the existence of a threshold the situation demands that one cannot cross despite having knowledge of the danger and the experience and skills of the individual). This is precisely what the stakes of risk-taking are. What is essential is to distinguish the risk on the basis on the individual's actual participation in his/her activity. The extent of his/her participation precisely underlines the concept of risk-taking that is inherent in risky behavior. Risk-taking may be defined as "the active participation of the individual in a behavior identified as potentially dangerous". It is the subjects themselves who seek out danger through certain behaviors that meet some of their needs. The Latin etymology of the word "danger", *dominarium, dominium*, means domination, possession, and power. Being in danger means being exposed to a threat (a person or a situation that may inflict physical or psychological harm). Seeking out danger "intentionally" is tantamount to claiming that one can dominate the threat instead of undergoing it. The motivational aspect is therefore an important dimension in decoding risk behavior. It is intentionality that defines risk-taking, just as in a real decision-making process. This implies a choice characterized by a degree of uncertainty as to the probability of success or failure (Michel, 2001a,b). The subject attributes a degree of usefulness to each probability (i.e., the benefit obtained from taking the risk). Danger is sought if it allows something and/or meets certain needs of the subject. This harks back to Bernoulli's theory of utility (1738) (i.e., the individual tries to maximize not the expected value of risk-taking but its utility). Various authors working on risk have incorporated this theory of utility into their thinking by conceptualizing risk-taking as a rational choice where the actions of the individual reflect his/her interests (Wilde, 1994). Thus, risk avoidance or engagement in a dangerous activity is primarily underpinned by a complex set of interactions between the individual and the environment that is mediated by the perceived benefits.

2.3 DEVELOPMENTAL APPROACH TO RISK-TAKING

From birth on, every human is faced with the issue of satisfying two conflicting needs that are part and parcel of the same process: seeking safety while and exploring one's environment, a form of exploration which is itself a source of risk. Every individual in his/her development has to manage both of these needs in order to be able to exist. Risk-taking is thus one of the basic elements involved in acquisition, independence, and individuation (Lerner & Tubman, 1991). For this reason, adolescence – a period of life when all these processes are fully active – is characterized as a time for taking on challenges, an age when the subject is most vulnerable to risk. This transitional period in life gives rise to experimentation with many behaviors, some of which are considered dangerous to one's health and welfare (Muus & Porton, 1998). Indeed, most risk-taking behaviors follow developmental patterns, their frequency and intensity increasing with age until adolescence and then decreasing (DiClement et al., 1996). In this perspective, risk-taking can be considered as being dependent on the psychobiological restructuring that takes place during adolescence (Michel, 2001a,b; Dahl & Spears 2004). Indeed, all corporal, morphological and physiological changes promote active behaviors. The youngster's increasing physical strength and heightened impulses lead him/her to try out new types of behavior that all inherently represent a danger owing to his/her inexperience. Risk-taking behavior may also allow the young person to explore his/her identity and to express his/her autonomy vis-à-vis other teenagers. Risk-taking seen as a social behavior (establishing an identity for the subject) could therefore be an effective means for some youngsters to gain independence and break away from their parents' control. By testing their courage to face up to danger, teenagers symbolically break the barriers of being children by becoming part of the peer group and associating risk-taking with one of the oldest traditions of humanity: rites of passage. Danger may thus be seen as a source of benefit for some young people and therefore as being even attractive.

2.3.1 Human factors: are we all equal?

The analysis of risk-taking behavior shows that while there are organizational and environmental factors of vulnerability, they are specific to individuals. Indeed, we are not all equal when it comes to risk-taking. For example, some psychological profiles (e.g., stress adjustment, personality profiles) seem to favor risk-taking behaviors more depending on the context, with consequences that may be harmful or beneficial to the individual (Michel & Purper-Ouakil, 2006).

2.3.2 Sensation-seekers

To a very large extent, risk-taking behaviors appear to be motivated by the search for intense emotions (i.e., an "adrenaline buzz"). This is what underlies a series of actions involving the speed, intoxication, and headiness that some people deliberately seek. This attraction for activities and behaviors inducing strong stimuli has been conceptualized as a dimension of personality by authors such as Zuckerman (1979) as "sensation-seeking" and Cloninger (1993) as "novelty-seeking". The quest for new intense forms of stimulation that may be conceptualized in psychobiological terms is defined by the need for a wide range of different and complex experiences and sensations that may lead the subject to engage in disinhibiting behaviors and risky physical and social activities.

Take sensation-seeking. It was in 1964 that Zuckerman, who was studying the individual valence of responses from subjects under conditions of isolation or sensory over-stimulation, used the concept of optimal level of stimulation in terms of activation based on the work of Hebb (1955) (see Figure 2.1; Zuckerman et al., 1964) to conceptualize his first theory of sensation-seeking. His first version of the theory posited that sensation-seekers function in a way as to maintain "an optimal high level of stimulation". According to Zuckerman, every human being has an optimal level of stimulation corresponding to his/her threshold. When we fall below our "threshold" (e.g., mild sensations induced in daily life), we experience boredom and disinterest. However, when our feelings exceed the threshold, we feel anguish, anxiety, or pain. Thus, individuals who have a very high threshold (high sensation-seekers) will be sensitive to the monotony of everyday life and tend to seek activities, behaviors or substances allowing them to feel sensations corresponding to their own personal threshold. Zuckerman, who later shifted toward a psychobiological model of sensation-seeking, changed the concept of an optimal level of activation to an optimal level of activity within the catecholaminergic system and particularly the dopaminergic system (Zuckerman, 1995).

The concept of sensation-seeking was operationalized by the creation of a scale, the sensation-seeking scale (SSS), in an adult version (Zuckerman, 1984) and a version for teenagers (Michel et al., 1999). Studies have

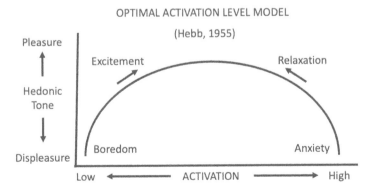

Figure 2.1 Model of optimal level of activation.

shown that subjects who are characterized by this personality profile may be described as excitable, impulsive, and sometimes disorganized. They quickly engage in new interests, seek thrill, and adventure and prefer activities without rules or constraints. Such individuals tend therefore to seek out the conditions in their environment that enable them to satisfy this need for stimulation. Indeed, this stimulation-seeking may be found in association with the use of psychoactive substances, dangerous behavior in motor vehicles such as speeding or not wearing a helmet, and in certain professions like mountain guide, racing car driver, and in some risky sports like base jumping and sky diving (for review, see Michel, 2001a,b). Without this desired and repeated exposure to stressful or anxiogenic situations, the individual becomes depressed. Adepts of this type of stimulation create a lifestyle based on intensity and novelty (c.f., professional activities, sports) in accordance with their personality. However, such risk-taking is favored even more by a psychological profile that itself is encouraged by a professional subculture that values it. Indeed, this propensity to seek our strong emotions is likely to be found in certain professionals who implicitly run risks (e.g., firefighters). However, in professions where this phenomenon is a central to the occupation, the fundamental issue is to establish the value and the need for risk-taking. At what point should the risk be taken? And is it sometimes more dangerous not to take it?

2.4 IS INFORMATION A PROTECTIVE FACTOR?

Do individuals who engage in hazardous activities perceive risks differently? Several authors have attempted to identify the various factors involved in risk assessment and their role in risk-taking. These factors include degree of personal control, experience of the practice, the effect of the group and information on the risks involved (Michel, 2001a,b). Other factors that influence the perception of risk include age and gender. Indeed, age appears to be a determining factor in the assessment of risk. On the one hand, some studies have shown that older people ascribe greater intensity to it and are thus more cautious. On the other hand, young people find it more difficult to weigh up the negative consequences of a behavior (Otani et al., 1992). For them, there is no underestimating the danger. Rather, in activities where a risk is involved, young people seek pleasure and thrills, with the result that their threshold of tolerance of the risk is pushed back, perhaps even leading them to refuse to obey the law that governs the activity. Indeed, prevention campaigns that focus on the effects and consequences associated with risky behaviors seem ineffective for some adolescents. Is information about risk a protective factor? Paradoxically, this even encourages some individuals in their quest of risk. Adolescents experiment with risky behaviors even though they are aware of the consequences associated with them. According to Tiemann and Tiemann (1983), the degree of risk amplifies the sense of anticipation of the benefits it may procure. The greater the risk, the greater the benefits will be seen as significant. For individuals with the sensation-seeking personality profile, information does not act as a protective factor. On the contrary, risk has a power of attraction, the dangerousness of the activity being a source of excitement and stimulation.

2.5 TOWARD AN INTERACTIONAL APPROACH TO RISK-TAKING

An interactionist approach to risk-taking attempts to link the major factors of vulnerability. Figure 2.2 shows the main human and environmental factors that may play a role in risk-taking depending on the nature of the event/situation. The model demonstrates the human (e.g., personality profiles, stress, psychopathologic state) and environmental (e.g., occupational and non-professional) factors that may lead the individual to undertake risky behavior. The factors should be analyzed from an interactionist perspective and not independently of each other. What makes individuals vulnerable is the nature of the relationship between these factors. Factors such as personality, psychopathology (human factors), cultural background, the organization of one's professional activity and life events such as previous

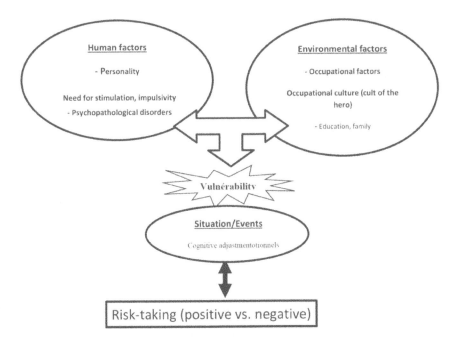

Figure 2.2 Model of human-environment interaction in risk-taking.

accidents (environmental factors) influence how an individual manages an event in terms of cognitive (e.g., perception of danger/benefits), emotional (e.g., stress management) and behavioral (e.g., anger, fatigue, substance abuse) adjustment. Thus, the interactionist approach looks at individuals not in terms of what they are but how they interact with the environment. From these variables, their role in taking a "reasoned" risk can be deduced. Preventive action in such cases could target both the professional and extra-professional environment of the individual, especially if the latter shows signs of indulging in risky behavior that may endanger him/her and others.

For this reason, the notion of "reasoned", "positive" risk-taking needs to be worked on. In the field of prevention, the emphasis is too often placed on eradicating the risk. Yet risk-taking is not only necessary in some situations; not taking a risk can sometimes be a source of danger.

2.6 DISCUSSION AND CONCLUSION

Risk-taking is an integral part of the development of every human being. To test and get to know oneself and to "exist", man must take risks of a physical, mental, and intellectual nature. Risk-taking is necessary not only in terms of ontogenetic development (human development) but also more widely in phylogenetic terms (development of the human species). Without risk-taking, progress in knowledge and technology is impossible. Without risk, no gain is possible. Individualization, civilization and the development of man and humanity inevitably imply risk-taking, provided that the risk taken is controlled and reasoned. The challenge in psychology is to differentiate the "pathogenic" risk-taking that characterizes several dangerous behaviors that have been identified in psychopathology and health (substance abuse, road risks, sporting, and occupational risks) from the "reasoned" risk-taking that may be particular to each individual but is mostly driven by very specific goals in terms of performance, competitiveness and objectives and in a wide range of sporting, professional and intellectual domains. Clearly, certain situations and environments favor the emergence of certain forms of risk behavior or risk-taking. The influence of the group and the pressure arising from interpersonal relations are important factors to be considered when analyzing the determinants of this type of conduct, especially since the cult of the feat or heroic behavior is strong in some walks of life. On the one hand, such praise of danger may lead some already predisposed individuals to take careless, thoughtless risks. On the other hand, this propensity to strong stimulation, boldness, or the need to undertake challenges that characterizes some individuals is, when appropriate, to be encouraged or controlled. The modern figure of the "hero" is more in the order of the fictional than the real. The self-denial whose virtues are extolled by the "hero", thereby bestowing on him the status of a superman, this ideal, may lead him to cast to the wind any precautionary approach in the quest for achievement. Therefore, the fundamental challenge is to shift from an ideology of heroism to a culture of "reasoned" risk-taking. In terms of human factors, we need to identify the specific characteristics of these "risk-takers" as potentially being skills to be recognized, analyzed, and optimized according to the circumstances so that the individual can in turn shift from the thoughtless and the dangerous to the reasoned. Longitudinal programs based on education about risk and danger could prove fruitful, whereas prevention campaigns often aim at eradicating risk. Education about risk could take the form of scenarios that aim at allowing young people not to give in to peer pressure but to better understand and regulate their emotions, to identify sufficiently stimulating environments for them so that they may satisfy their need for stimulation while developing their skills and personal resources. Risk-taking is not only necessary in some situations and areas: its absence can also be a source of danger and lead to a lack of empowerment or personal achievement. The challenge facing us is to move from a culture of danger and insecurity to one of education about "reasoned" risk-taking, the impact of which could engender essential benefits for future generations.

NOTE

1 Exceeding one's physical limitations induces biological changes, including the secretion of endorphins, which are endogenous drugs that may lead to the onset of a certain degree of dependence (Calhoon, 1991).

REFERENCES

Bernoulli D. (1954). Exposition of a new theory of risk evaluation. Econometrica, 22, pp. 23–36.

Calhoon L.L. (1991). Sensation seeking, exercise, and dopamine beta hydroxylase. *Personality and Individual Differences*, 12, pp. 903–907.

Cloninger C.R., Svrakic D.M. & Przybeck T.R. (1993). A psychological model of temperament and character. *Archives of General Psychiatry*, 50, pp. 975–990.

Dahl R.E. & Spears L.P. (2004). *Adolescent Brain Development: Vulnerabilities and Opportunities*. New York: New York Academy of Sciences.

DiClement R.J., Hansen W.B. & Ponton L.E. (1996). Adolescent at risk. In R.J. DiClement, W.B. Hansen & L.E. Ponton (Eds.), *Handbook of Adolescent Health Risk Behavior*. New York, NY: Plenum.

Hebb D.O. (1955). Drives and the CNS, conceptual nervous system. *Psychological Reviews*, 62, pp. 243–254.

Kouabenan D.R., Cadet B., Hermand D. & Muñoz Sastre M.T. (2006). *Psychologie du risque: identifier, évaluer, prévenir*. Bruxelles: De Boeck.

Lerner R.M. & Tubman J.G. (1991). Developmental contextualism and the study of early adolescent development. In R. Cohen & A.W. Siegel (Eds.), *Context and Development*. Hillsdale, NJ: Laurence Erlbaum Associates.

Michel G., Mouren-Siméoni M.C., Perez-Diaz F., Carton S. & Jouvent R. (1999). Validation and construction of sensation seeking scale for adolescent. *Personality and Individual Differences*, 26, pp. 159–174.

Michel G. (2001a). Recherche de sensations et sur-éveil corporel à l'adolescence. *Neuropsychiatrie de l'Enfance et de l'Adolescence*, 49, pp. 244–251.

Michel G. (2001b). *La prise de risque à l'adolescence: pratique sportive et usage de substances psychoactives*. Paris: Masson Collection: Les âges de la vie.

Michel G. & Purper-Ouakil D. (2006). *Personnalité et développement: Du normal au pathologique*. Paris: Dunod.

Michel G., Bernadet S., Aubron V. & Cazenave N. (2010). Des conduites à risques aux assuétudes comportementales: le trouble addictif au danger. *Psychologie française*, 55(4), pp. 341–353.

Muus R.E. & Porton H D. (1998). Increasing risk behavior among adolescents. In R.E. Muss & H.D. Porton (Eds.), *Adolescent Behavior and Society* (5th ed.) (pp. 422–431). Boston, MA: McGraw-Hill.

Otani H., Leonard S.D. & Ashford V.L. et al. (1992). Age differences in perception of risk. *Perceptual and Motor Skills*, 74, pp. 587–594.

Tiemann A.R. & Tiemann J.J. (1983). Cognitive map of risk and benefit perceptions. In The Annual Meeting of the Society for Risk Analysis, New York.

Turner C., McClure R. & Pirozzo S. (2004). Injury and risk-taking behavior – a systematic review. *Accident Analysis and Prevention*, 36, pp. 93–101.

Wilde G.J. (1994). *Target Risk*. Toronto, Ontario: P D E Publications.

Zuckerman M. (1979). Sensation seeking and risk-taking. In C.E. Izard (Ed.) *Emotions in Personality and Psychopathy* (pp. 163–197). New York, NY: Plenum Press.

Zuckerman M. (1984). Sensation seeking: a comparative approach to a human trait. *Behavioral and Brain Sciences*, 7, pp. 413–471.

Zuckerman M. (1995). Good and bad humors: biochemical bases of personality and its disorders. *Psychological Science*, 6, pp. 325–332.

Zuckerman M., Kolin E.A., Price L. et al. (1964). Development of a sensation-seeking scale. *Journal of Consulting and Clinical Psychology*, 28, pp. 477–482.

Chapter 3

Decisions and risks manifesting themselves through movements performed by workers

Sylvie Leclercq
INRS, French Research and Safety Institute for the
Prevention of Occupational Accidents and Diseases
Vandoeuvre-lès-Nancy, France

CONTENTS

3.1 INTRODUCTION

In the working world, decisions relating to the design of work situations or preventive approaches, for example, are taken at different levels in the company, or even beyond. These decisions determine the risks presented by any given work situation. In addition, decisions made by the workers contribute to whether the risks manifest themselves or not. From this point of view, all

DOI: 10.1201/9781003221609-3

these decisions can be considered individual and/or collective risk-taking. Thus, risk-taking, the subject of this volume will be addressed in this chapter through the concept of decision-making. We will show the role played by the risk characteristics and the objectives pursued in making decisions.

First, we focus on occupational risks presenting risks of occupational accidents (OAs) and occupational diseases (ODs) based on their indicators of frequency and severity (Table 3.1). We then set out characteristics shared

Table 3.1 Characterization of occupational accidents from the statistical data for 2012 available from the CNAMTS (French National Health Insurance Fund for Salaried Workers)[2]

	Injury caused by an element with which any contact or proximity lead to an injury (high voltage source, chemical product, moving part of a machine etc.)	Injury caused by an element with which the worker interacts regularly without being injured (floor, wall, static part of a machine, carried item etc.)		
Proportion among all OA with sick leave	30%	70%		
Sub-categories		Falls from a **height**[a] (4%)	Accidents during **manual handling** (34%)	Trips, slips, jamming, collisions, and other **movement disturbances** and also **pains** which occur when the person is in motion (32%)
Companies involved	Mainly companies in the **secondary sector**	**All companies**		
Severity in terms of percentage • PI[b] • Days lost due to TI[c] • Deaths	30% (PI) 26% (TI) 81% (death)	70% (PI) 74% (TI) 19% (death)		
Decrease in IdF[d] between 1998 and 2012 (%)	**9.8**			
		5.7		

a   Falls in situations involving work at a height, i.e., situations resulting from the site of work (rooftops, walkways, beams, etc.) or the use of specific equipment (ladders, scaffolding, work platforms).
b   Permanent incapacity.
c   Temporary incapacity.
d   Index of frequency = (number of OA requiring sick-leave × 1000)/size of the working population.

by most risks of OA/OD which are risks manifesting themselves through workers' movements. Then we present the effect of these characteristics on risk perception and thus on the related decisions. These characteristics will sometimes be compared to the characteristics of industrial risks,[1] which are other risks present in some sociotechnical systems.

Second, we address the sociotechnical system which must deal with risks of OA/OD and industrial risks to consider the distinct objectives pursued during management of each type of risk. These different objectives give rise to decisions which will represent a tradeoff and will determine the risks encountered in each working situation. We then present the objectives pursued in an actual working situation which give rise to decisions which will also be trade-offs.

Finally, based on these elements, prevention of OA with movement disturbance (OAMD) will be addressed, which are particularly difficult for preventers to deal with. The nature of the hazard and how the risk is perceived are obstacles to prevention of these accidents. More specifically, they can contribute to behaviors qualified as risk-taking. Whether the risk of OAMD is considered during the design phase is discussed, either as part of the design of working situations or the design of preventive actions.

In some cases, the discussion is illustrated using data relating to the risks of OA/OD as a whole. On numerous other points, given the author's experience, it focuses on data relating to some of the most frequent types of OA: OAMD.

3.2 DECISIONS AND CHARACTERISTICS OF RISKS OF OA/OD

3.2.1 Frequency and severity of risks of OA/OD

In 2007, the International Labor Organization (ILO, 2007) estimated that every year, around two million people worldwide die because of OA/OD, and that in over 200 countries around 270 and 160 million people are victims of OA and OD, respectively. In the same year, and according to the European agency for safety and health at work, 2.9% of workers had an accident resulting in more than three days' sick leave (EASHW, 2013). In 2005, 84,070 OD were recognized in 12 European Union member states (the 15 member states except Germany, Greece, and Ireland – EASHW, 2010) – a 32% increase relative to 2002. In France, the national statistics for OD for companies in the general regime (CNAMTS, 2015) recorded 36,871 OD in 2004. Their number increased continuously up to 2011, when 55,057 OD were recognized. It subsequently decreased slightly (51,631 OD in 2014) and has now stabilized. The CNAMTS figures also show that the number of OA decreased significantly between the end of the 1970s and the end of the 1980s, even though the number of workers remained relatively stable. Thus, over one million OA requiring sick leaves were recorded in 1977 in companies in the general regime. In 1987, this number was 662,800. Since then, the drop has been less pronounced, with 618,263 OA requiring sick leave recorded in 2013. Based on the evolution of various indicators, the risk of OA, unlike the risk of OD for example, appears to have been controlled for several decades. However, the results in terms of severity and in particular the number of working days lost due to OA, are less favorable: more than 28 million working days were lost in 1977, compared to more than 37 million in 2012.

3.2.2 Manifestation of risks of OA/OD and concepts of hazard/exposure

Analysis of the databases listing OA/OD reveals that:

- Most occupational risks manifest themselves through workers' movements (Leclercq et al., 2015). In France, around 80% of recognized ODs are musculoskeletal disorders (MSD) and 70% of OA are falls from a height, accidents during manual handling or accidents caused by slips, trips, collisions, or any other movement disturbance.
- The other risks (chemical risk, electrocution, injury during contact with moving parts of a machine, occupational cancer, noise-induced hearing loss, etc.) introduce an element external to the worker into the injury production, which obviously has a negative impact when close to or in contact with the worker (high-voltage source, carcinogen, etc.). This element corresponds to the concept of hazard when representing "dangerous" situations or situations "presenting a risk".

These observations first question the hazard concept when the risks manifest themselves through workers' movements, which is the case in the majority of OA/OD. In these cases, the injury is not caused by something which is by its nature incompatible with the individual's physical integrity (see Dumaine 1985 for example in relation to the phenomenon of accident), rather it is the consequence of constraints exerted on the musculoskeletal system beyond its functional capacities. When a collision occurs, for example, an element external to the victim contributes to the injury. This may be a wall if an arm is bumped against it. In most cases, this element is not identified on principle as harmful or likely to contribute to causing injury or damage.

It is important to underline this characteristic of most risks of OA/OD in terms of hazards as it determines how the risk is perceived, and consequently it influences the related decisions (see Section 3.2.4). It also determines the possible preventive actions (see Section 3.4.1).

The notion of exposure, itself indissociable from the concept of hazard, is thus called into question. About the risk of slips, trips, jamming or any other movement disturbance during activity, it is essential to consider that on principle (1) it is present whenever movement is involved in the work activity and (2) that any element in the work environment can contribute to causing injury if an impact against this element occurs when movement is disturbed.

3.2.3 Diffuse nature of the risk of OA/OD reinforced by how it is recorded

Risks of OA/OD, by their very definition, affect workers in their occupations. Insurance companies record OA and OD in databases with a view to compensation. Each time a worker is victim of an occupational injury, the event must be recorded. Numerous accidents which occur in the workplace involve a single victim. The risk is thus diffused. Even when an accident affects several workers it will be recorded as many "accidents" as there are victims. Similarly, an OD must be recorded every time a worker develops one. When several workers develop an OD following the same acute accidental exposure to a toxic product for example, as many declarations of OD will be made as there are affected workers.

In contrast, an industrial accident can affect workers, those availing of the service rendered, the environment and/or material damage to company installations. It will be recorded as a single event in the databases set up to provide returns on experience (e.g., Accident Analysis, Research, and Information [ARIA] base developed by the Bureau for Analysis of Risks and Industrial Pollution which reports to the French Minister for ecology, sustainable development, and energy). The same accident will also be recorded as many "occupational accidents" as there are victims among workers.

Thus, how OA/ODs are recorded reinforces the diffuse nature of the associated risks. It is also interesting to note that the literature sometimes uses the terminology "personal accident" to evoke OA, to distinguish them from "process accident" (Hopkins, 2009; Kjellén, 2009).

3.2.4 How risks manifesting themselves through workers' movements are perceived

A report published by the European Agency for Safety & Health at Work in 2012 (EASHW, 2012) dealt with the perception of risks linked to nanomaterials and retrieved a certain number of characteristics of a risk influencing their perception and the level of concern that it raises, from Slovic (2000). It appears that most risks of OA/OD, given the risks manifesting themselves through workers' movements, present several characteristics which decrease the perceived risk. These characteristics for the risks presented in EASHW (2012) are the following:

- diffuse nature (see Section 3.2.3);
- whose manifestations are familiar, historical (case of MSD and accidents caused by movement disturbance);

- people have more or less learned to live with the risk, which does not provoke terror or fear (see Section 3.2.2);
- people believe they have (some degree of) control over the risk;
- associated with exposure that appears to be voluntary;
- by their nature, known to those who are exposed;
- explained by science, at least partially;
- risks which manifest themselves suddenly or cause reversible damage;
- rarely cause death (although it does sometimes occur).

These characteristics influencing how the risk is perceived will, with the indicators of frequency and severity, influence decisions determining the risks in working situations and their prevention. Other elements also affect these decisions, in particular the objectives pursued. They are addressed in the following paragraph.

3.3 DECISIONS AND OBJECTIVES PURSUED

All sociotechnical systems must involve means to prevent risks of OA/OD. Some systems must also deal with industrial risks: for example, installations classed as representing an environmental hazard, transport of hazardous matter or distribution and use of domestic gas. Let's consider a company which must manage both types of safety and examine some objectives pursued at different levels.

3.3.1 Decisions and objectives when managing safety

The decisions made to manage safety in a sociotechnical system depend on the objective pursued. However, as underlined by Leclercq et al. (2018), these will differ depending on the type of risk. In the case of industrial risk, it is avoiding loss of control in a production process, leading to damage having broad consequences, affecting the surrounding populations, the company environment and installations, and the operators involved in the process. When dealing with a risk of OA/OD, the objective is to avoid bodily harm to operators who contribute directly or indirectly to how this type of process proceeds. How do these different objectives interact when making decisions relating to "safeties" management?

In addition, the indicators in terms of occupational safety may provide no information on the process safety. This was revealed by the analysis of the explosion which occurred at the "BP Texas City Plant" in 2005. Indeed, Grote (2012) reported that this analysis indicated a high level of occupational safety, but that less attention was paid to indicators relating to the process safety. These observations raise several questions, and in particular the following: what decisions led to this situation? What indicators/observations were used? Do the actions to prevent one type of risk involve (or not) the prevention of another type of risk?

3.3.2 Decisions in working situations

Occupational risks manifest themselves in work situations, where the worker applies controls through their activity to ensure safe production while also preserving their own health and safety. Some ergonomic models, used in the field of occupational health and safety, focus on these controls. One of the oldest is the 5-square diagram (Leplat & Cuny, 1974). More recently, Vézina (2001) developed a comparable model to analyze the development of MSD. This model formalizes movement during an activity and the possible effects on the risk of developing MSD of the controls made by the workers through their movements. In the context of the prevention of industrial risks, Daniellou et al. (2010) also used a model centered on the worker's activity, based on the following three objectives pursued: production, occupational health and safety, and industrial safety. The model represented in Figure 3.1 was inspired by these three models. It represents

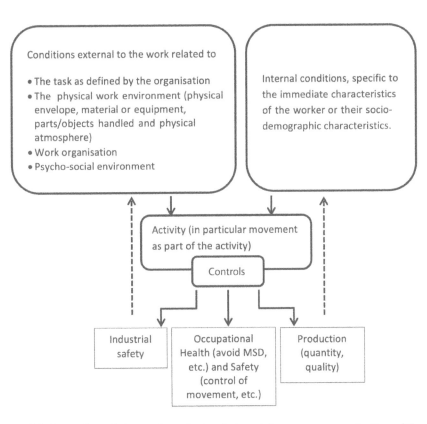

Figure 3.1 Controls made by workers during their activity to ensure production while preserving their health and safety. This diagram was inspired by the 5-square diagram. (Leplat & Cuny, 1974), Vézina's model (2001) and the model presented by Daniellou et al. (2010.)

the controls made by the worker and their possible effects on the risks of OA/OD, industrial risks, and production. As risks manifesting themselves through workers' movement, whether OD or OA risks (see Table 3.1), are the most frequent, Figure 3.1 formalizes movement as part of the activity.

Those workers make trade-offs between production and safety has long been known. For example, Faverge (1970) reported controls made in accident-prone situations, in particular involving co-activity, recovery when technical incident for example or resume after recovery.

The extracts from descriptions of accidents and recurrent accident scenarios due to a movement disturbance and presented below (accident 1 and scenario 1) reveal trade-offs between production in a broad sense and the worker's safety.

Accident 1. Extract of the description contained in the EPICEA database (EPICEA, 2011). Upon arrival at the building site, a reinforced fire-door was found to have been deposited against a wall by the structural work team. No workers from the structural work team were present on the site, so the manager and his worker decided to move the fire-door outside together to place it in the garden. As they were moving it toward the garden, passing through the French window, the worker stumbled against the bottom stop limit of the rolling shutter causing him to lose his balance. As he fell, without releasing the door, the worker's right hand was trapped between the door and the cement floor.

Scenario 1. Extracted from Leclercq and Thouy (2004). At the start of the day, the worker received a list of times for meetings scheduled with clients. Walking between the worker's vehicle and the client's home was made difficult due to environmental conditions – snow had fallen, making the ground slippery and masking obstacles. These difficulties were combined with increased journey times when travelling by road because of the snow. The requirement to respect the schedule for the meetings led to worker's movement disturbance when he was alone and could make up for "time lost" when driving.

Trade-offs between industrial safety and occupational health/safety adopted by workers are less well known. Indeed, a worker's activity can have direct or indirect consequences on

- their own health and safety;
- and/or that of their colleagues who are sometimes subordinates;

- and/or that of clients availing of a service rendered when the company provides a service;
- and/or the company's installations;
- and/or the environment.

Recurrent scenario 2 presented below reveals a trade-off made by a worker between their own safety and the safety of passengers (availing of the service rendered).

Scenario 2. Extracted from Leclercq et al. (2007). This recurrent accident affected several ticket inspectors on trains when the train was departing the station. The carriages of the trains involved had a step and a bar outside, and the system to open/close the doors was often defective. Passengers running late therefore often attempted to board trains even after the signal indicating departure of the train. After having whistled the signal, the inspector was walking toward the train while also monitoring passengers' movements to prevent them from taking risks by attempting to board the train once it had started moving. In one case, an inspector stumbled against the tactile ground surface indicator, another bumped into a post on the platform, a third missed their step when boarding the train.

The worker is thus obliged, at certain times, to make trade-offs which can appear to be "risk-taking", but which emerge as a consequence of the conditions in which the activity is performed, and the objectives pursued.

3.4 PREVENTION AND RISKS OF ACCIDENTS WITH MOVEMENT DISTURBANCE (AMD)

According to Rasmussen (1997), "In any well-designed work system, numerous precautions are taken to protect the workers against occupational risk and the system against major accidents, using a 'defense-in-depth' design strategy".

Among the risks of OA manifesting themselves through workers' movements (see Table 3.1), those over which the preventer (and the designer?) is the most helpless is the risk of AMD, i.e., the risk of slips, trips, collisions, jamming, and other movement disturbances except for those leading to falls from a situation at height. These accidents represent one-third of OA and have been defined in operational terms (Leclercq et al., 2010). In the following paragraphs, we present some elements relating to their prevention, at the time of design.

3.4.1 Nature of the hazard: an obstacle to setting up barriers

It is difficult for preventers and designers to anticipate the risks of OAMD, because these risks cannot be apprehended from an obvious hazard (element with which any contact or proximity leads to injury or disease – see Section 3.2.2). Indeed, risks related to obvious hazards are the subject of regulations with which preventers and designers must comply. For example, barriers to prevent contact or proximity with the hazard may be required. Haddon's strategies (see Figure 3.2) aiming to prevent any contact or proximity between an exposed target and a hazard are generally not practically applicable when dealing with risks of OAMD. Take the case of a nurse whose arm collided with a cupboard as she was moving a patient (see Leclercq et al., 2014). Among the two elements which produced the injury, the energy of movement of the arm (the hazard according to the energy model applied in Figure 3.2) is "borne" by the victim. As a result, several of Haddon's strategies cannot be applied. The actions relating to the cupboard are limited (e.g., reduction of sharp edges) and often aim to reduce the severity of the injury, but not to prevent the movement disturbance itself.

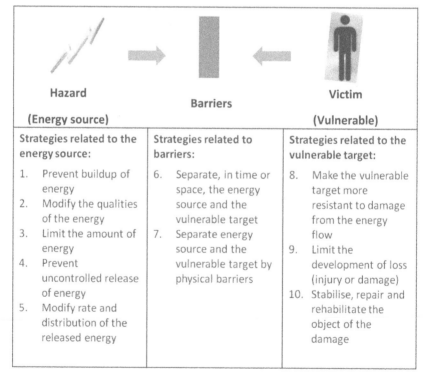

Hazard (Energy source)	Barriers	Victim (Vulnerable)
Strategies related to the energy source:	Strategies related to barriers:	Strategies related to the vulnerable target:
1. Prevent buildup of energy 2. Modify the qualities of the energy 3. Limit the amount of energy 4. Prevent uncontrolled release of energy 5. Modify rate and distribution of the released energy	6. Separate, in time or space, the energy source and the vulnerable target 7. Separate energy source and the vulnerable target by physical barriers	8. Make the vulnerable target more resistant to damage from the energy flow 9. Limit the development of loss (injury or damage) 10. Stabilise, repair and rehabilitate the object of the damage

Figure 3.2 Haddon's ten accident prevention strategies. (Taken from Kjellén 2000 who adapted it from Haddon 1980.)

3.4.2 Risk perception: an obstacle to prevention

The sinister nature of the risk of OAMD is eloquent. Preventive studies and practices are not aligned with the risks, partly because the representations and perception of these risks are an obstacle to prevention. Indeed, this risk presents several characteristics that reduce the perceived risk (see Section 3.2.4). In addition, in many cases these accidents are considered benign and banal everyday occurrences. The victim is the first to be blamed by themselves and by others. As a result, this type of accident is rarely analyzed, which contributes to maintaining this causal explanation. However, the risk of OAMD increases or decreases depending on the occupational conditions and thus the conditions in which movements are performed at work. Studies performing a more extensive analysis of these accidents indicate that the risk factors are linked to all the components of the working situation: use of equipment (Kines, 2003), configuration of access systems (Leclercq et al., 2007), design of working situations (Derosier et al., 2008), work organization (Leclercq, 2014), and safety management (Bentley & Haslam, 2001).

3.4.3 A look at design from reports of accidents

3.4.3.1 Design of work situations

No element in the labor code explicitly relates to OAMD. However, some requirements related to the design of workplaces (characteristics of the buildings, lighting, safety training, etc.) can contribute to their prevention. Nevertheless, these very general rules do not consider activity performance and remain mostly insufficient, as shown by the two examples of accidents presented in Section 3.3.2. and the questions they raise relating to design. At a specific moment in time, a tactile ground surface indicator (scenario 2) and the limit of the rolling shutter (accident 1) represented an obstacle to movement during the work activity, and this obstacle could not be adequately accounted for by the worker at that time. These elements (tactile ground surface indicator and rolling shutter limit) are positioned during the design phase, in some cases to prevent another risk (the tactile ground surface indicator prevents the risk of falling onto the tracks, for sight-impaired people) and in other cases to meet product specifications (the rolling shutter limit guides the rolling shutter as it closes). Could the accidental situations evoked have been anticipated at the design stage? It is difficult to know. However, we can hypothesize that the guiding factors during design, were that the step, the post, the tactile ground surface indicator, or the rolling shutter limit be perfectly visible, there is no reason that people, and even more so those habitually working in the environment would fail to take them into consideration during their movements. Nevertheless, the accidents presented above show that at certain times it can be very difficult for a worker to fulfill their task while also controlling their movement.

3.4.3.2 Design of prevention

The most difficult OAMD to anticipate are those which result from the presence at a given time of an unusual combination of usual conditions. The difference from a usual situation is thus like that revealed by "emerging" industrial accidents that result from variability in usual conditions (Hollnagel, 2004). It was in response to the emerging nature of industrial risks that Hollnagel et al. (2006) introduced the concept of resilience. This concept aims to reinforce the adaptive and anticipatory capacities of the system to respond to the risks remaining despite the implementation of preventive barriers. Thus, prevention must be designed based on an increase in the adaptive and anticipative capacities of the system to help prevent OAMD, for which preventive safety barriers are often, not insufficient but impossible to implement in practice (see Section 3.4.1 and Leclercq et al., 2013).

3.5 CONCLUSION

Most risks of OA/OD present numerous specificities that reduce the perceived risk. In a sociotechnical system, these risks are combined with other risks that are managed with a view to meeting distinct objectives, in particular industrial risk and the risk of nonconformity with the planned production. Trade-offs are thus made, at different levels, leading to the presence of "residual" risks in a working situation, where the risks manifest themselves.

Management of occupational risk relies more particularly on work situation design and on how the work is organized. The "residual" risk in a working situation must be partly managed by the worker, and their contribution is particularly important to cope with the risk of OAMD.

NOTES

1 The terminology qualifying this type of risk is rich and varied: industrial risks but also major technological risks or process accident risk.
2 Table 3.1 was adapted by the author of the chapter. It illustrates how these conclusions were reached from data relating to OA in France. More than 640,000 OA requiring sick leave were recorded in 2012 for a population of around 19 million workers.

REFERENCES

Bentley, T. A., & Haslam, R. A. (2001). Identification of risk factors and countermeasures for slip, trip and fall accidents during the delivery of mail. *Applied Ergonomics*, 32(2), 127–134. doi: 10.1016/S0003-6870(00)00048-X.
CNAMTS (2015). *Les chiffres de la sinistralité en 2014 et faits marquants dans les secteurs d'activité*. Press release, November.

Daniellou, F., Simard, M., & Boissières, I. (2010). *Facteurs humains et organisationnels de la sécurité industrielle: un état de l'art (Vol. Cahiers de la sécurité industrielle)*. Toulouse, France: Institut pour une culture de sécurité industrielle.

Derosier, C., Leclercq, S., Rabardel, P., & Langa, P. (2008). Studying work practices: a key factor in understanding accidents on the level triggered by a balance disturbance. *Ergonomics*, 51(12), 1926–1943. doi: 10.1080/00140130802567061.

Dumaine, J. (1985). La modélisation du phénomène accident. *Sécurité et Médecine du Travail*, 71, 11–22.

EASHW (European Agency for Safety and Health at Work) (2010). *OSH in figures: work-related musculoskeletal disorders in the EU – facts and figures*. Luxembourg: Publications Office of the European Union. doi: 10.2802/10952.

EASHW (European Agency for Safety and Health at Work) (2012). *Risk perception and risk communication with regard to nanomaterials in the workplace*. Luxembourg: Publications Office of the European Union. doi: 10.2802/93075.

EASHW (European Agency for Safety and Health at Work) (2013). *Estimating the cost of accidents and ill health at work*. Luxembourg: Publications Office of the European Union. doi: 10.2802/8236.

EPICEA (2011). [Online] Available at: http://www.Inrs.fr/accueil/produits/bdd/epicea.html.

Faverge, J. M. (1970). Plenary session: the operator's reliability and safety in industry L'Homme Agent d'infiabilite et de Fiabilite du Processus Industriel. *Ergonomics*, 13(3), 301–327.

Grote, G. (2012). Safety management in different high-risk domains – all the same? *Safety Science*, 50(10), 1983–1992. doi: 10.1016/j.ssci.2011.07.017.

Haddon, W. (1980). The basic strategies for preventing damage from hazards of all kinds. *Hazard Prevention*, 16, 8–12.

Hollnagel, E. (2004). *Barriers and accident prevention, or how to improve safety by understanding the nature of accidents rather than finding their causes*. Hampshire, United Kingdom, Ashgate.

Hollnagel, E., Woods, D. D., & Levenson, N. (2006). *Resilience engineering: concepts and precepts*. Hampshire, United Kingdom: Ashgate Publishing Limited.

Hopkins, A. (2009). Thinking about process safety indicators. *Safety Science*, 47(4), 460–465. doi: 10.1016/j.ssci.2007.12.006.

ILO (2007). *Project on economic dynamics of international labour standards*. Retrieved 17-02-2017, from http://www.ilo.org/wcmsp5/groups/public/—ed_norm/-relconf/documents/meetingdocument/wcms_084831.pdf.

Kines, P. (2003). Case studies of occupational falls from heights: cognition and behavior in context. *Journal of Safety Research*, 34(3), 263–271. doi: 10.1016/S0022-4375(03)00023-9.

Kjellén, U. (2000). *Prevention of accidents through experience feedback*. London, United Kingdom: CRC Press (Taylor and Francis Group).

Kjellén, U. (2009). The safety measurement problem revisited. *Safety Science*, 47(4), 486–489. doi: 10.1016/j.ssci.2008.07.023.

Leclercq, S. (2014). Organisational factors of occupational accidents with movement disturbance (OAMD) and prevention. *Industrial Health*, 52(5), 393–398. doi: 10.2486/indhealth.2014-0076.

Leclercq, S., Cuny-Guerrier, A., Gaudez, C., & Aublet-Cuvelier, A. (2015). Similarities between work related musculoskeletal disorders and slips, trips and falls. *Ergonomics*, 58(10), 1624–1636. doi: 10.1080/00140139.2015.1031191.

Leclercq, S., Monteau, M., & Cuny, X. (2010). Avancée dans la prévention des "chutesde plain-pied" au travail. Proposition de définition opérationnelle d'une nouvelle classe: "les accidents avec perturbation du mouvement (APM)". *Perspectives interdisciplinaires sur le travail et la santé*, 12(3), 155–166.

Leclercq, S., Monteau, M., & Cuny, X. (2013). Quels modèles pour prévenir les accidents du travail d'aujourd'hui? [Theories and methodologies]. *Le Travail Humain*, 76(2), 105–127.

Leclercq, S., Morel, G., & Chauvin, C. (2018). Process versus personal accidents within sociotechnical systems: loss of control of process versus personal energy? *Safety Science*, 102, 60–67. http://dx.doi.org/j.ssci.2017.10.003.

Leclercq, S., Saurel, D., Cuny, X., & Monteau, M. (2014). Research into cases of slips, collisions and other movement disturbances occurring in work situations in a hospital environment. *Safety Science*, 68, 204–211. doi: 10.1016/j.ssci.2014.04.009.

Leclercq, S., & Thouy, S. (2004). Systemic analysis of so-called 'accidents on the level' in a multi trade company. *Ergonomics*, 47(12), 1282–1300. doi: 10.1080/00140130410001712627.

Leclercq, S., Thouy, S., & Rossignol, E. (2007). Progress in understanding processes underlying occupational accidents on the level based on case studies. *Ergonomics*, 50(1), 59–79. doi: 10.1080/00140130600980862.

Leplat, J., & Cuny, X. (1974). *Les accidents du travail*. Paris: Presses Universitaires de France.

Rasmussen, J. (1997). Risk management in a dynamic society: a modelling problem. *Safety Science*, 27(2–3), 183–213. doi: 10.1016/S0925-7535(97)00052-0.

Slovic, P. (2000). *The perception of risk*. London; Sterling, VA: Earthscan Publications.

Vézina, N. (2001, 2001/10/3–5). La pratique de l'ergonomie face aux TMS: ouverture à l'interdisciplinarité. Paper presented at the SELF-ACE conference. Ergonomics for changing work, Montreal, Canada.

Chapter 4

The EAST 'broken-links' method for examining risk in sociotechnical systems

Neville A. Stanton
University of Southampton
Southampton, United Kingdom

Catherine Harvey
University of Nottingham
Nottingham, United Kingdom

CONTENTS

4.1 INTRODUCTION

The term 'Sociotechnical Systems' (STSs) is used to refer to the interaction between humans and machines, from the small and simple to the large and highly complex (Walker et al., 2008, 2010b; Read et al., 2015). These subsystems operate and are managed as independently functioning (autonomous) entities, with their own goals, but must collaborate with other subsystems to achieve the higher goals of the STS (Dul et al., 2012; Wilson, 2012). A key characteristic is that these goals can only be achieved by the STS and not by individual subsystems functioning in isolation (Rasmussen, 1997; von Bertalanffy, 1950). STS present unique challenges for safety management

DOI: 10.1201/9781003221609-4

and risk assessment (Rasmussen, 1997; Alexander and Kelly, 2013; Flach et al., 2015; Waterson et al., 2015). Traditional approaches to risk assessment, such as HAZOP, THERP, HEIST and SHERPA, are typically reductionistic in nature (Stanton et al., 2013), focusing on individual tasks and technologies rather than the system as a whole (Stanton, 2006; Stanton and Stevenage, 1998; Stanton et al., 2009; Waterson et al., 2015). These methods use error taxonomies to identify risk, but recent research has suggested that the term 'human error' is obsolete (Dekker, 2014). In its place the term 'human performance variability' has been proposed, which includes both normative and nonnormative performance. This latter approach emphasizes the broad spectrum of human behavior, rather than a dichotomy, and therefore a need to build resilient systems (Hollnagel et al., 2006). The systemic accident analysis (SAA) approach treats systems as whole entities with complex, nonlinear networks (Underwood and Waterson, 2013). A number of SAA methods were assessed for their potential for prospective risk analysis within STS in a previous study (Stanton et al., 2012). Some system methods incorporate error taxonomies, such as CREAM, HFACS and STAMP, which, given recent shift away from the term of 'human error', is something of a conundrum. Rather than considering risks in systems to be the result of error, the approach taken in this paper is to propose risks as the failure to communicate information via social and task networks. This type of failure may be seen in several major incidents. For example, in the MS Herald of Free Enterprise accident (1987), the state of the bow doors was not communicated to the ships bridge (Noyes and Stanton, 1997). So the ship left harbor and subsequently capsized. In the Kegworth air disaster (1989), the aircraft failed to communicate which engine was on fire, leading to the pilots shutting down the wrong engine (Griffin et al., 2010; Plant and Stanton, 2012). In the Ladbroke Grove Rail incident (1999) the signals failed to communicate to the train driver that the section of the rail network was protected (Moray et al., 2016). Rather than stopping, the driver increased his speed as he passed the red signal leading to a collision with an oncoming high-speed intercity train (Stanton and Walker, 2011). So rather than conceiving these behaviors as errors, we have reconceived them as the failure to communicate information in the system. To analyze the information communications in systems, the Event Analysis of Systemic Teamwork (EAST) method was selected. EAST takes a different approach to the error taxonomic methods, by modeling and analyzing STS-level interactions. In a previous study, Stanton (2014) analyzed communications between various actors within a submarine control room: in contrast, this case study analyzes a retrospective account of actions within a Royal Navy training activity and is conducted at the macro level (Grote et al., 2014). The aim of this paper is to extend the EAST network-level analysis to include risk prediction by 'breaking' links within networks.

The EAST method was first proposed by Stanton et al. (2005, 2013) and was further elaborated by Stanton et al. (2008) for modeling-distributed cognition in STS. The method represents distributed cognition in networks that enables both qualitative and quantitative investigations to be performed (Stanton, 2014). One of the main advantages of EAST is its aim to capture the whole system, as opposed to reductionist methods which split a system into constituent parts for analysis (Walker et al., 2010a). It is therefore considered in this study to be a suitable technique for representing an STS and potential nonnormative behaviors. The analysis describes a system as three different types of network:

- Social: representing the agents (human, technical and organizational) within a system and communications between them,
- Task: representing the activities performed by the system and the relationships between them,
- Information: representing the information that is used and communicated within a system and links between different information types.

The social, task and information networks are developed individually and then combined to create a complete social-task-information network diagram, showing all the links and information flows (i.e., distributed cognition) within a network-of-networks. EAST has been applied in many domains, including aviation (Stewart et al., 2008; Walker et al., 2010a), military (Stanton et al., 2006; Stanton, 2014), road (Salmon et al., 2014), rail (Walker et al., 2006) and the emergency services (Houghton et al., 2006). The aim of this work is to extend the EAST method to consider risk in systems via a case study and provide an initial STS method and evaluation criteria presented by Harvey and Stanton (2014). The premise of the risk assessment is that STS failures are predominately caused by the failure to communicate information between agents and tasks. This will be studied within the context of the following case study.

4.2 CASE STUDY OF HAWK MISSILE SIMULATION TRAINING

Operation of the Hawk jet to simulate missile attacks against surface ships by the UK Royal Navy was selected as the case study. This activity is viewed as an STS because it comprises many interconnected subsystems, which are themselves complex. The context for this study is illustrated in an AcciMap (Rasmussen, 1997; Jenkins et al., 2010) in Figure 4.1.

The AcciMap places different subsystems within the STS at different levels and shows the links in communication and decision-making between the subsystems. Each node in the AcciMap is labeled (a, b, c...) to correspond with the description in the text in the following paragraphs. The year in which each event occurred is also included where applicable, to give an indication of timescale.

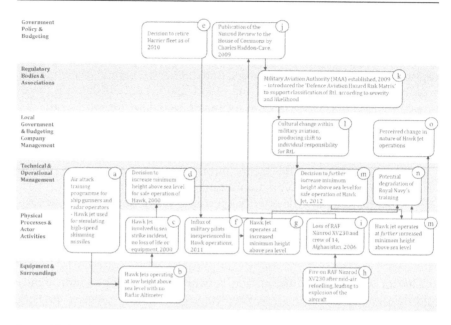

Figure 4.1 AcciMap showing subsystems within the Hawk Jet STS. (Please note that the labels do not indicate a timeline, rather they are added for clarity of the description below.)

The Royal Navy uses the Hawk Jet to simulate air attacks on ships during sea training of ships' gunners and radar operators (event 'a' in Figure 4.1). The Hawks are used to simulate enemy aircraft attacks and high-speed skimming missiles fired against ships (Royal Navy, 2012). To perform these simulation activities, the Hawk must be flown at a low height above sea level (b); however, the Hawk is not equipped with a Radar Altimeter (Rad-Alt) that provides a highly accurate measure of the altitude of the aircraft above the sea. This makes flying the Hawk accurately at very low levels extremely difficult and requires a high level of expertise to perform safely. Prompted by events over several years, there have been some significant changes to the method for assessing the safety of the Hawk STS.

In 2000, a Hawk Jet was involved in a sea strike incident because of very low-level flight (c). Although there was no resulting loss of life, this incident prompted a decision by the Royal Navy to increase the minimum allowable flying height above sea level for the Hawk (d).

As part of its Strategic Defence and Security Review, H M Government (MAA, 2010) took the decision to retire the Harrier jet from service in October 2010 (e). As a result of this, several Royal Navy pilots who would have flown the Harrier were diverted into the Hawk program (f). Traditionally the Hawk has been flown by civilian pilots under contract to the Royal Navy (2012): these pilots have extensive military experience in fast jets, which includes low-level flight supported by a Rad Alt. This experience provided mitigation against the risk to life (RtL) for the Hawk air attack simulation task; however, the cohort of military pilots did not have this same level of experience and the RtL had to be reassessed considering this (g).

In 2006, RAF Nimrod XV230 suffered a catastrophic explosion after a mid-air refueling procedure (h): this caused the deaths of all 12 crew members plus two mission specialists and total loss of the aircraft (i). The Government requested a comprehensive review into the airworthiness and safe operation of the Nimrod (j), which was delivered by Haddon-Cave (2009). The report described the development of the safety case for the Nimrod as 'a story of incompetence, complacency, and cynicism' (p. 161) and concluded that it was undermined by the widespread assumption that the Nimrod was safe because it had been flown successfully for the preceding 30 years (Haddon-Cave, 2009). The report also identified organizational changes in the years prior to the Nimrod accident as having significant influence; these included a shift in organizational culture toward business and financial targets 'at the expense of functional values such as safety and airworthiness' (Haddon-Cave, 2009, p. 355). As a consequence of the findings, Haddon-Cave (2009) recommended the establishment of an *independent* Military Aviation Authority (MAA) to properly assess RtL and shape future safety culture (k). Further recommendations included the need for strong *leadership*, a greater focus on *people* to deliver 'high standards of safety and airworthiness' (Haddon-Cave, 2009, p. 355) and increased *simplicity* of rules and regulations. The tragic consequences of the Nimrod accident, along with the recommendations of the Haddon-Cave report, effected a culture change within military aviation: this resulted in a decision to assign *individual* accountability for RtL assessments to 'Duty Holders' (DH), where previous responsibility for risk had been held at the organization level (l). The newly established MAA produced guidelines for the assessment of RtL, in the form of the Defence Aviation Hazard Risk Matrix (MAA, 2011) that supports the classification of single risks according to their estimated severity (catastrophic, critical, major, minor) and likelihood (frequent, occasional, remote, improbable). The resulting risk level determines at which level of DH the risk is held.

The organizational changes brought about by the events described above (i.e., influx of junior pilots) prompted reassessment of the RtL for the Hawk air attack simulation activity. The goal of safety management in the UK military is to reduce risk to a level which is As Low As Reasonably Practicable (ALARP): this is reached when 'the cost of further reduction is grossly disproportionate to the benefits of risk reduction'. The RtL for all Hawk operations is frequently reassessed and the shift in pilot experience levels, as described above, prompted changes to the RtL for the Hawk air attack simulation activity. To reduce this RtL to a level that was ALARP, a decision was taken by the Royal Navy DH with subject matter expert (SME) advice to further increase the minimum height above sea level (m).

A potential consequence of this decision is the degradation of Royal Navy surface fleet training against very live low-level targets, as the Hawk can no longer accurately simulate sea skimming missile attacks on surface ships (n). These events have changed the nature of Hawk operations within the UK MoD (o).

Potential risks to the safe operation of the missile simulation activity are, in part, assessed according to the MAA's Regulatory Articles (RAs) (MAA, 2011). This assessment is based on the principle that risks can be tolerated provided they are reduced to ALARP. The MAA regulatory policy outlined its approach to the management of RtL:

'Aviation DHs [Duty Holders] are bound to reduce the RtL within their AoR [Area of Responsibility] to at least tolerable and ALARP; the application of effective and coherent risk management processes will be fundamental to achieving this' (MAA, 2011, p. 18).

RA 1210 – Management of Operating Risk (Risk to Life) – defined risk as: 'a measure of exposure to possible loss [combining] severity of loss (how bad) and the likelihood of suffering that loss (how often)' (MAA, 2011, p. 1).

The MAA suggested that risks can be identified via several different methods including previous occurrences, checklists, HAZOPS, zonal hazard (safety) analyses and error trend monitoring. Previous work has showed that these techniques are likely to be inadequate for the analysis of STS (Stanton et al., 2012). RA 1210 specifically encourages the use of fault trees as accident models 'to assist understanding of the interrelationship between risks and to support the prioritization of effort to maximize safety benefit' (MAA, 2010, p. 6). This technique, along with other traditional error and risk prediction methods, does not account for the interactions of distributed actors within an STS (Salmon et al., 2011b). Furthermore, there is also no clear method outlined by the MAA for structuring risk identification; for example, the recommendation is that a combination of these methods should be used with the aim of identifying all credible risks, but there is no way of knowing when all possible credible risks have been defined and therefore how many methods to use and when to stop applying them.

The Hawk RtL case study was identified through interviews with an SME as part of this project. The analysts were provided with a high-level overview of the case study in an initial interview with the SME. This was followed up by a second, in-depth interview about the case study with the SME, conducted by two analysts. This resulted in a detailed account of the Hawk-Frigate STS, which was supplemented by extra information from official documentation including MAA (2010, 2011) guidelines, the official report into the Nimrod accident (Haddon-Cave, 2009) and Royal Navy safety assessment guidance (Royal Navy, 2012). The EAST method (Stanton, 2014) was used to develop the three network diagrams, based upon the analysis of all case study information, in an iterative process that involved the SME providing feedback during development.

4.3 ANALYSIS OF NETWORKS

Social Network Analysis (SNA) metrics provide quantitative measures that represent the structures and relations between nodes in the EAST networks (Baber et al., 2013, Driskell and Mullen, 2005, Walker et al., 2009). The SNA metrics describe individual nodes (including reception, emission, eccentricity, sociometric status, centrality, closeness, farness and betweenness). The SNA metrics applied in the current study, along with their descriptions, are presented in Table 4.1.

Analysis software, AGNA version 2.1 (Benta, 2005), was used to calculate the SNA metrics. For each EAST network, key nodes were identified according to sociometric status. Sociometric status was selected to define key nodes because it identifies the prominence of an individual node's communications with the rest of the network, which influences the whole network's performance (Stanton, 2014). In an STS, all of the nodes will have complex safety management rules and behaviors; however, as the 'key' nodes have the largest number of connections to the rest of the network, these nodes will have the highest degree of influence over the behavior of the entire STS. Sociometric status key nodes are defined as nodes that have a higher sociometric status score than the sum of the mean sociometric

Table 4.1 SNA Metrics, along with their descriptions

	Safety constraint	Description
Node-level metrics	Emission	The number of edges (links to other nodes) originating at that node
	Reception	The number of edges incident to that node
	Sociometric status	Number of communications received and emitted relative to the number of nodes in the network
	Bavelas-Leavitt (B-L) centrality	Degree of connectivity to other nodes in the network
	Eccentricity	Length of the longest geodesic path originating in that node (a geodesic path is between two given nodes that has the shortest possible length)
	Closeness	Inverse of the sum of the geodesic distances from that node to all the other nodes, i.e., extent to which a node is close to all other nodes
	Standard closeness	Closeness multiplied by (g-1), where g is the number of nodes in the network
	Farness	Sum of the geodesic distances from that node to all other nodes
	Betweenness	Frequency with which a node falls between pairs of other nodes in the network

status score plus the standard deviation sociometric status score for all nodes in the network. SNA metrics were calculated for the EAST networks created for this case study: key agents for sociometric status are indicated in the social, task and information network diagrams below.

4.4 RESULTS

The step-by-step application of the shortened version of EAST is described in detail in the following sections. This is accompanied by the outputs of the method along with interpretation of the results. The first stage in EAST was the identification of all social, task and information nodes within the Hawk missile simulation case study, based on the SME's account of activities, which informed the analysts' knowledge of the case study. The nodes were arranged in social, task and information networks and links drawn between related nodes. Related nodes were those between which some information was transferred. As well as providing a visual representation of an STS, the EAST network diagrams can be analyzed to produce quantitative SNA metrics.

4.4.1 Social network

Seven social 'agents' and their connections were identified from the Hawk RtL case study with the SME: these are shown in Figure 4.2. The social network was constructed by first identifying the main agents that are in the system, then by examining the interdependencies between those agents. The SME agreed that the social network was a reasonable representation of the main agents and their relationships. The 'edges', or links between the agents show where information is transferred and the direction of transfer. There are 19 edges in total; in some cases, information transfer is reciprocal but in others it only goes in one direction between two agents. The 'pilot' node was identified as the key agent according to sociometric status (1.33): it has the highest number of links to and from other nodes in the network, in fact the pilot receives and/or emits information from/to all the other agents in the social network.

The pilot had the highest betweenness score (10.0) as it is located on the paths between several other agent pairs. The pilot also had the highest score for reception (6), highlighting a high degree of connectivity to other nodes in the network and indicating that the pilot's actions and communications are integral to the functioning of the STS. The high farness score for the regulator (10) indicates that this agent is located furthest from most other nodes, and this is supported by the information in the case study, which showed that the regulator only communicates with the DH and possibly the pilots but has no contact with the frigate or crew. This is because the regulator in this case is the MAA that does not have direct control over the Navy's surface ship operations. The sea scored highest for emission (4) and lowest for reception (0) as

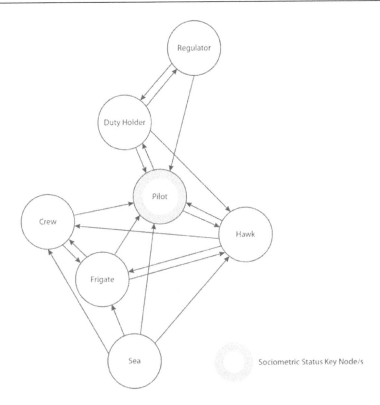

Figure 4.2 Social nodes and their links within the STS.

it does not receive information from any other nodes but is used for feedback only. In this sense the sea can be regarded as a 'passive' agent, as it cannot respond to feedback; the social agents can only respond to it.

4.4.2 Task network

Ten task nodes and their connections were identified from the Hawk RtL case study with the SME: these are shown in Figure 4.3. The task network was constructed by first identifying the main tasks that are performed by the system, then by examining the interdependencies between the tasks. The SME agreed that the task network was a reasonable representation of the system. There are 12 edges in total and in all cases the transfer is unidirectional.

In the task network, key nodes were identified as 'safe control of aircraft to simulate missile' and 'issuing of RtL document' that had sociometric status scores of .56 and .44, respectively. The task network contains more nodes but fewer edges than the social network, indicating that there are fewer communications between tasks. Cohesion is zero because there are no mutual, or bidirectional, links between nodes. The highest score for betweenness was

Figure 4.3 Task nodes and their links within the STS.

for 'safe control of aircraft to simulate missile' (45) demonstrating that this task is integral in the STS as it is located between a high number of other task nodes. This is unsurprising as this task can be the main objective of the STS configuration investigated in this case study. This task also scored highest on emission (2), reception (3) and B-L centrality (6.26), as well as sociometric status (.56), showing a high level of connectivity to other nodes.

4.4.3 Information network

EAST identified 25 information nodes and their connections were based on the Hawk RtL case study and further knowledge of the STS from the SME: these are shown in Figure 4.4. There are 50 edges in total; however, in this case, the links are not directional.

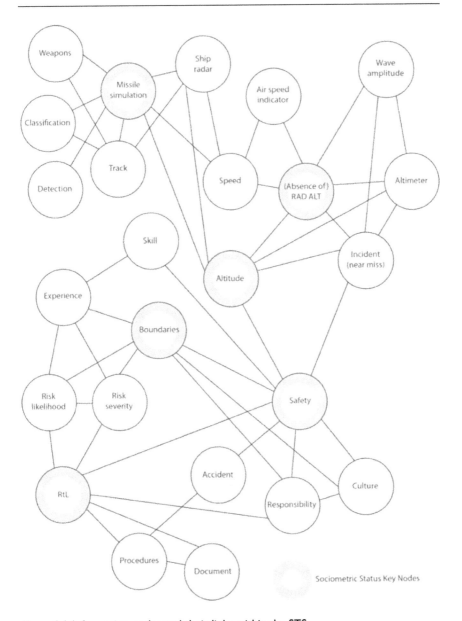

Figure 4.4 Information nodes and their links within the STS.

Six information nodes were identified as key nodes according to socio-metric status: missile simulation, (absence of) Rad Alt, altitude, boundaries, safety and RtL. Safety had the highest betweenness (315.8), standard closeness (.53) and B-L centrality scores (18.51) and this reflects the importance of this in the case study: the aim within the STS is to achieve a safe solution for missile simulation. Density and cohesion were relatively low, i.e., compared with the social network, as the edges between nodes were single and nondirectional.

4.4.4 Broken-Links analysis

Studies of networks have discussed the effects of removing one or more nodes from a network on the resilience of that network to systemic failures and the resulting destabilization (Baber et al., 2013; Houghton et al., 2008; Stanton, 2014). This has been used to explore the resulting influence on network structure, rather than as a method for predicting specific risks. Previously, the network diagrams in EAST have been used to provide a visual representation of a system to further the users' understanding of distributed cognition (Stanton, 2014). In this study however, the EAST network analysis was extended to identify and examine possible risks by 'breaking' the links between the various nodes, in a similar approach to the removal of nodes, to explore system effects.

Broken-links represent failures in communication and information transfer between nodes in the networks and these failures can then be used to make predictions about the possible risks within the STS. Previously, 'broken-links' have only been investigated by EAST analysts when looking retrospectively at accidents to identify underlying causes. Griffin et al. (2010) demonstrated that the broken-link between the Engine Vibration Indicator and the pilots in the cockpit was a causal factor in their failure to shut down the correct engine in the Kegworth accident. If this information had been communicated more effectively it could have helped to prevent the crash. Similarly, the EAST method has been adapted to analyze incidents of fratricide (Rafferty et al., 2012) although this has been conducted as retrospective and concurrent, rather than predictive, analyses. The broken-links analysis was performed for the Hawk missile simulation case study, on the social and task networks shown in Figures 4.2 and 4.3, respectively. The information network was not subject to the broken-links analysis because broken-links between information nodes were not considered to represent risks, as they are caused by a failure in either the social or task networks. In other words, information does not fail in isolation; it is the failure to use or communicate the information correctly and in all cases, this can be attributed to social nodes, task nodes or both. For the social and task networks, each link was identified and documented in a table. The combined EAST networks diagram (see Figure 4.5) shows the information network tagged with the social networks nodes (to show who owns each information node in the network) and grouped by the task network nodes (to show which task each information node belongs to). Details on construction of the combined network have been reported by Stanton (2014). This network was used to identify what information (from the information network) should be communicated from the origin node to the destination node in the task

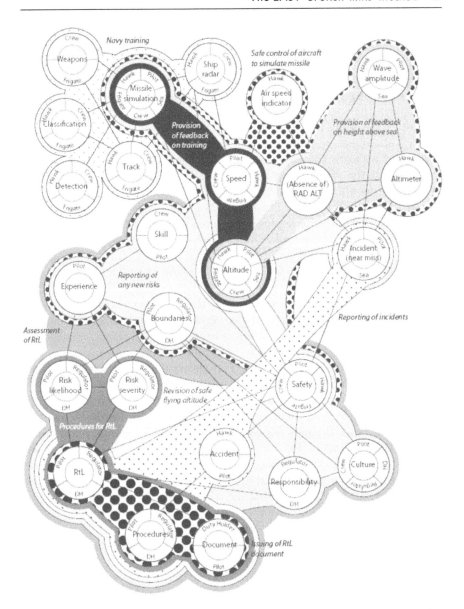

Figure 4.5 Combined information-task-social network for the Hawk case study (shading represents the different tasks being undertaken).

and social networks and therefore what information would not be communicated if the links between the nodes in the task and social networks were removed.

Figure 4.5 also shows the combined information-task-social network as a single depiction of the entire STS. This shows the overlaps between the three networks, in other words, which information is being communicated by which agents in which tasks, and how these nodes are interlinked.

In order to conduct the broken-links analysis, the social and task networks were compared to the combined information-task-social network in turn. For example, there is a reciprocal relationship between the DH and the pilot (in the social network shown in Figure 4.2) and the DH and the pilot share the nodes of boundaries, RtL, risk likelihood, risk severity, procedures, document, responsibility and safety (in the combined information-task-social network shown in Figure 4.5). The risk assessment procedure requires that the relationship between the DH and pilot be interrogated to see what would happen if each information element was not transmitted, as shown in Table 4.2. The pilot was identified as having the highest Sociometric Status in the analysis presented in Figure 4.2, so was chosen for the illustration of the broken-links analysis in Table 4.2. Although the pilot is linked to all other agents in the social network, for the purpose of illustration just their reciprocal relationship with the DH is presented in Table 4.2.

Table 4.2 shows the risks resulting from the failure to pass relevant information between DH and pilot and vice versa. Anecdotal evidence from our SME suggests that there is variability in what individual pilots will chose to report back to the DH, as they have different interpretations of what they consider to be a risk and near miss. This shows that there is at least some face validity for the approach we have proposed.

In the similar manner to the social-information broken-links analysis shown in Table 4.2, there is a task-information broken-links analysis in Table 4.3. From the task network, there is a unidirectional relationship between the 'Issuing of RtL document' and the 'Revision of safe flying altitude' (in the task network shown in Figure 4.3) and they overlap in the combined information-task-social network (shown in Figure 4.5). The risk assessment procedure requires that the relationship between the 'Issuing of RtL document' and the 'Revision of safe flying altitude' be interrogated to see what would happen if each information element was not transmitted, as shown in Table 4.3. The 'Issuing of RtL document' was chosen as it has the highest Sociometric Status in the analysis presented in Figure 4.3. The 'Safe control of aircraft to simulate missile' was chosen for the same reason and is paired with 'Navy training' for the purposes of offering an illustrative example of the method in Figure 4.3.

Examination of the analysis in Tables 4.2 and 4.3 offers a systematic approach for examining a system of operation in a holistic manner. For example, increasing the safe flying attitude (see Figure 4.1) has led to the altitude profile of the Hawk not matching that of the low flying missile (see Table 4.3). This has meant that reducing the risk for the Hawk pilots has had a negative effect on training of the crew on the Frigate, ultimately increasing their risk. So while the top of Table 4.3 is about improving the safety of the pilot, by increasing altitude for example, the bottom of

Table 4.2 Extract from broken-links analysis for EAST social network

From (agent)	To (agent)	Information not communicated	Resulting risk	Mitigation strategy
Duty Holder	Pilot	Boundaries	Pilots are not aware of the boundaries for flight operations and for the identification and reporting of risks within this	Boundaries for risk reporting must be made clear to pilots as part of the RtL process
Duty Holder	Pilot	RtL	Pilots are not made aware of the results and consequences of the RtL assessment process after it is conducted at DH level	Results and consequences of the RtL assessment process must be effectively communicated to pilots
Duty Holder	Pilot	Risk likelihood	Pilots are not made aware of risks assessed that their likelihood of occurrence	Risks identified as having a high likelihood of occurrence must be reported to pilots
Duty Holder	Pilot	Risk severity	Pilots are not made aware of risks assessed and their severity of impact	Risks deemed as having a high severity of impact must be reported to pilots
Duty Holder	Pilot	Procedures	Pilots are not aware of how the RtL process is conducted at DH level and of procedures for reporting incidents to the DH	Pilots must be provided with clear procedures describing the assessment of RtL at DH level and the reporting of risks to DH
Duty Holder	Pilot	Document	Pilots are not provided with documentation covering the RtL process and its results	Pilots must be provided with documentation covering the RtL process and its results
Duty Holder	Pilot	Responsibility	Pilots are not aware of the DH's nor their own responsibilities for safety	The responsibilities of both the pilot and DH for safety must be clearly defined and understood by pilots
Duty Holder	Pilot	Safety	Pilots do not receive information about the safety of operations, based on the RtL assessment process	The safety of operations, as assessed during the RtL process, must be reported to the pilots
Pilot	Duty Holder	RtL	The DH does not receive information about new risks identified by the pilots	Pilots must clearly report all relevant risks to the DH

(Continued)

Table 4.2 Extract from broken-links analysis for EAST social network *(Continued)*

From (agent)	To (agent)	Information not communicated	Resulting risk	Mitigation strategy
Pilot	Duty Holder	Risk likelihood	The DH does not receive information about the likelihood of new risks identified by the pilots	Pilots must report their estimate of the likelihood of occurrence of all relevant risks
Pilot	Duty Holder	Risk severity	The DH does not receive information about the severity of new risks identified by the pilots	Pilots must report their estimate of the severity of impact of all relevant risks
Pilot	Duty Holder	Incident (near miss)	The DH does not receive information about incidents (or near misses) which occur during Hawk operations	Pilots must clearly report all relevant incidents which occur during Hawk operations to the DH
Pilot	Duty Holder	Experience	The DH cannot learn from the pilots' experience of Hawk operations and the risks encountered	Pilots must clearly report their experience levels to the DH. Pilots must report their assessment of risks and any consequent assumptions, based on this experience
Pilot	Duty Holder	Skill	The DH cannot learn from the pilots' skill in Hawk operations	Pilots must clearly report their skill levels to the DH. Pilots must report their assessment of risks and any consequent assumptions, based on this skill level
Pilot	Duty Holder	Safety	The DH does not receive information about the safety of Hawk operations	Pilots must report their estimates of the safety impact of any risks identified in Hawk operations to the DH
Pilot	Duty Holder	Culture	The DH is not aware of the culture of safety among the Hawk pilots	Pilots must consider and report the estimated influence of safety culture on the risks to Hawk operations

Table 4.3 Extract from broken-links analysis for EAST task network

From (task)	To (task)	Information not communicated	Resulting risk	Mitigation strategy
Issuing of RtL document	Revision of safe flying altitude	Document	The information contained in the RtL document does not trigger a revision of safe flying altitude	The RtL document must be used by regulators to inform changes to regulations and safety guidance where appropriate
Issuing of RtL document	Revision of safe flying altitude	RtL	The outcome of the RtL process outlined in the RtL document does not trigger a revision of safe flying altitude	The outcomes of RtL assessment must be used by regulators to inform changes to regulations and safety guidelines where appropriate
Issuing of RtL document	Revision of safe flying altitude	Risk likelihood	The outcome of the Risk likelihood assessment, conducted as part of the RtL process and outlined in the RtL document, does not trigger a revision of safe flying altitude	The outcome of the Risk likelihood assessment, conducted as part of the RtL process and outlined in the RtL document, must be used to inform changes to regulations and safety guidelines where appropriate
Issuing of RtL document	Revision of safe flying altitude	Risk severity	The outcome of the Risk severity assessment, conducted as part of the RtL process and outlined in the RtL document, does not trigger a revision of safe flying altitude	The outcome of the Risk severity assessment, conducted as part of the RtL process and outlined in the RtL document, must be used to inform changes to regulations and safety guidelines where appropriate
Issuing of RtL document	Revision of safe flying altitude	Safety	The safety implications of the RtL process outlined in the RtL document do not trigger a revision of safe flying altitude	The safety implications of RtL assessment must be used by regulators to inform changes to regulations and safety guidelines where appropriate

(Continued)

Table 4.3 Extract from broken-links analysis for EAST task network *(Continued)*

From (task)	To (task)	Information not communicated	Resulting risk	Mitigation strategy
Issuing of RtL document	Revision of safe flying altitude	Responsibility	Responsibility for the revision of safe flying altitude is not outlined in the RtL document	Responsibility for changes to regulations and safety guidelines based on RtL assessment must be clearly assigned and accepted
Safe control of aircraft to simulate missile	Navy training	Missile simulation	The overall control of the Hawk does not adequately simulate missile attack on the frigate to aid with training	The operation of the Hawk must aid Navy training for missile attack situations
Safe control of aircraft to simulate missile	Navy training	Speed	The speed profile of the Hawk does not adequately simulate missile attack on the frigate to aid with training	The speed of the Hawk during missile simulation must be sufficiently realistic to aid Navy training for missile attack situations
Safe control of aircraft to simulate missile	Navy training	Altitude	The altitude profile of the Hawk does not adequately simulate missile attack on the frigate to aid with training	The altitude of the Hawk during missile simulation must be sufficiently realistic to aid Navy training for missile attack situations
Safe control of aircraft to simulate missile	Navy training	Track	The track of the Hawk does not adequately simulate missile attack on the frigate to aid with training	The track of the Hawk during missile simulation must be sufficiently realistic to aid Navy training for missile attack situations

table three shows that this could reduce the safety of the Navy frigate crew as they do not receive realistic training. The benefit of systems approaches is that the knock-on effects become more readily apparent.

4.5 DISCUSSION

This work aimed to explore the use of a modified version of EAST (network modeling and broken-links analysis, see Stanton, 2014) in a case study of a Royal Navy training activity. First, the findings of this study are discussed in terms of twelve criteria that were identified as essential for methods designed to analyze the human component of STS (Stanton et al., 2012; Harvey and Stanton, 2014). This enabled comparisons to be made between the method and the current RtL procedure used in the Hawk missile simulation case study. Second, the modifications and extensions to EAST are discussed with reference to use of the method as an assessment of potential risks within an STS.

Aviation accidents, as with most accidents in STS, usually occur due to a conjunction of factors (Hodgson et al., 2013; Jenkins et al., 2010) and it is therefore essential that analysis methods are able to explore all of these factors by taking an integrated and holistic approach (Ramos et al., 2012; Salmon et al., 2011b). EAST specifically enables the exploration of the social, task and information components of the STS, allowing a high-level model of the STS to be created (visual diagrams) and analyzed (social network metrics). This visual component is likely to help analysts and other stakeholders to understand the interactions within networks: this is an advantage over many other methods such as HAZOP, Fault Tree analysis, as well as the MAA's RtL/HRM approach. Baber et al. (2013) argued that it is sensible to speak of a 'useful' (rather than 'complete') network, as there will always be a possibility that some connections have been left out due to not being observed, reported and/or documented. This is certainly applicable to the networks generated by EAST, as it is impossible to know whether an analysis has been exhaustive, and it is therefore sensible to assume that it has not. It is also particularly true in this case as the analysis was performed on an SME's reports of activities within the STS, rather than communications between STS actors (as in Stanton, 2014). A consequence of this approach is a lack of richness of information, although if the main contribution of EAST lies in its ability to visually represent an STS, then this may not be a significant issue. EAST includes the calculation of SNA metrics that provide the analyst with quantitative values to represent various characteristics of the networks. In this way, the analysis encompasses all elements of a STS and provides the analyst with an understanding of the structure of a system as a whole and the relationships between individual system components. These metrics can provide potential insight into the resilience of the networks (Stanton, Harris and Starr, 2016).

The inclusion of particular agents in an accident model is dependent on the information put into the analysis and therefore on the analysts and SMEs involved. This is true for more traditional HAZOP and error identification methods and the current RtL assessment process, as well as EAST. However, because HAZOP and RtL assessment essentially focuses on a list of potential errors, there is no formal procedure for identifying the decision makers involved. In contrast, EAST enabled a visual representation of the decision maker agents and their relationships with other nodes in an STS to be constructed, thereby encompassing the identification of decision makers into the analysis process. This can allow analysts to understand where responsibility for risks resides within the STS and so target mitigation strategies appropriately (Lundberg et al., 2010). This case study showed that EAST provided a useful visual representation of relationships between the various components of the STS. EAST examines the links between nodes and so is focused on communications, and therefore on the consequences of an action at a node, rather than its causes (Rafferty et al., 2012; Walker et al., 2010a).

In this case study, the analysts used a modified version of EAST, concentrating on the social, information and task networks (Stanton, 2014). Guidance is provided on structuring a model of the system/STS under investigation and there are numerous examples of previous EAST models (e.g., Griffin et al., 2010; Rafferty et al, 2012; Walker et al., 2010a; Stanton, 2014) in the literature. The 'break-link' process is very straight forward indeed and would be a useful addition to the current RtL assessment process (Haddon-Cave, 2009). The guidance states that risks should be identified from several sources including HAZOP, error data and experience of previous events; however, there are no explicit instructions on how many of these methods to use and when to stop this analysis. This means that the RtL assessment may proceed without a comprehensive list of potential risks. It appears that EAST could be a useful model for ensuring that this does not happen; however, it is important to note that provision of guidance may not be sufficient for successful application of STS methods. The training requirements of these methods can often be high for practitioners, with many citing a lack of time and difficulty accessing new information as barriers to STS analysis (Underwood and Waterson, 2013).

Stanton et al. (2012, 2014) previously suggested that EAST could be suitable for prospective analysis of STS risks, however these studies only demonstrated the utility of methods for retrospective analysis (Salmon et al., 2011a,b; Waterson et al., 2015). In this study, EAST has been applied to an STS that is currently in operation in order to investigate the ability of methods to model the future state of an STS. The Hawk missile simulation STS has already experienced and been impacted by incidents (e.g., Hawk sea strike) and accidents (e.g. Nimrod), but this analysis focused on the prediction of a future state given the changes in the STS, such as the alteration in safe flying altitude for the Hawk and the effects of this on missile simulation for the frigate and crew. Having said this, the emphasis with EAST is not on predicting accidents; rather, it is about creating a comprehensive model of the links and information flows within the STS and by doing so making the analysts aware of potential breakdowns and failures that may occur in the future. This means that the success of EAST for prospective analysis is dependent on the participation in the assessment process of those who will be impacted by these failures and those that can apply the appropriate mitigation strategies.

In summary, this study used a modified version of EAST, following the examples in Stanton (2014). In this case only the network analysis phases of EAST were applied (followed by a new phase: broken-links analysis that has not been previously reported) because the preliminary stages of

EAST were negated by having already collected and represented the data via interviews with an SME. Furthermore, some of the EAST methods require communications data that was lacking in this particular case study as the information came from an SME's account of the STS. Compared to Stanton's (2014) analysis of the operations within a submarine control room, the current study analyzed activity at a more macro level, using an SME's account of activities within the STS rather than a transcript of direct communications between STS actors. Recording and transcribing communications within a working system in real time is difficult, time-consuming and potentially disruptive to the STS under investigation. The approach presented in the current paper would be easier for personnel within the STS itself to apply, to support their own safety management and risk prediction activities, as it relies on a macro-level account of actions and relationships with an STS. This also allows these personnel to create a systems view of the STS of interest, which, as previously discussed, offers benefits over traditional analysis techniques. However, the absence of communications data obviously means that the analysis lacks detail and a richness of information that comes from speech data. This also meant that frequency of communications could not be represented in the same way as Stanton (2014). So in effect, the network diagrams in this case study offer a basic visual representation of an STS which could aid understanding of the relationships between agents, tasks and information as well as their combination. Of greater importance is the extension to EAST presented in this paper: the broken-links analysis. To identify potential risk in the STS, these links between nodes in the networks had to be examined in more detail; this was accomplished in the broken-links analysis. In this phase, each link between the task and social nodes was 'broken' to illustrate the effect of a communication breakdown between nodes. In this case study, 19 social links and 12 task links were broken and assessed against numerous information nodes, resulting in the identification of 137 risks in total. These breakdowns would result in a failure in information transfer, so each broken link was analyzed against the information nodes to identify potential risks. This extension to the EAST method provides a structured method for identifying all of the risks within an STS. The broken links can be listed in table form, along with 'to' and 'from' information detailing the origin and destination nodes between which information is transferred. The broken-links analysis is concluded by developing mitigation strategies for each of the identified risks, in a similar way to other error analysis methods.

4.6 CONCLUSIONS

This paper presents a case study of the extension to the EAST method applied in the analysis of an STS, specifically Hawk missile simulation to aid with training of Navy crew. The approach models the STS in two stages.

The first stage is to model the system as three networks (social, information and task) and the second stage is to break the links in the social and task networks to discover what risks are introduced by the failure to communicate information. This approach is based on the premise that most, if not all, accidents and near misses are caused, at least in part, by the failure to communicate information between agents and tasks. By enabling the generation of a system-model, EAST ensures that all of the components of interest within an STS have at been identified and this should lead to a more comprehensive analysis of potential risks. The extension to EAST offers a holistic, structured and systematic approach to the identification of information communication failures in task and social networks. This is a radical departure to the taxonomic approaches traditionally used to model risk in systems. The approach can be applied to any STS in any domain where an EAST model has been constructed. Future work could explore the risks associated with multiple communication failures occurring simultaneously as well as considering the degree of resilience in the network models.

ACKNOWLEDGEMENTS

The authors would like to thank Wing Commander Neil Bing (Bingo) of Air Cap SO1 Lightning, RAF High Wycombe, for his account of the Hawk Risk to Life case study and his very valuable insights into the challenges faced by this complex Socio-Technical System. This work was part-funded by the Defence Human Capability Science and Technology Centre (DHCSTC) grant reference TIN 2.002.

REFERENCES

Alexander, R. & Kelly, T. (2013). Supporting systems of systems hazard analysis using multi-agent simulation. Safety Science, 51, 302–318.

Baber, C., Stanton, N. A., Atkinson, J., McMaster, R. & Houghton, R. J. (2013). Using social network analysis and agent-based modelling to explore information flow using common operational pictures for maritime search and rescue operations. Ergonomics, 56, 889–905.

Benta, M. I. (2005). Studying communication networks with AGNA 2.1. Cognition, Brain, Behavior, 9, 567–574.

Dekker, S. (2014). The field guide to understanding 'human error'. Aldershot: Ashgate.

Driskell, J.E. & Mullen, B. (2005). Social network analysis. In: Stanton, N. A., Hedge, A., Brookhuis, K., Salas, E. & Hendrick, H. (eds.) Handbook of human factors and ergonomics methods. Boca Raton, FL: CRC Press.

Dul, J., Bruder, R., Buckle, P., Carayon, P., Falzon, P., Marras, W. S., Wilson, J. R. & Van der Doelen, B. (2012). A strategy for human factors/ergonomics: developing the discipline and profession. Ergonomics, 55, 377–395.

Flach, J. M., Carroll, J. S., Dainoff, M. J. & Hamilton, W. I. (2015). Striving for safety: communicating and deciding in sociotechnical systems, Ergonomics, 58 (4), 615–634.

Griffin, T.G.C., Young, M. S. & Stanton, N. A. (2010). Investigating accident causation through information network modelling. Ergonomics, 53, 198–210.

Grote, G., Weyer, J. & Stanton, N. A. (2014). Beyond human-centered automation – concepts for human–machine interaction in multi-layered networks. Ergonomics, 57 (3), 289–294.

Haddon-Cave, C. (2009). The Nimrod review. An independent review into broader issues surrounding the loss of the RAF Nimrod MR2 aircraft XV230 in Afghanistan in 2006. London: The Stationery Office.

Hodgson, A., Siemieniuch, C. E. & Hubbard, E.-M. (2013). Culture and the safety of complex automated sociotechnical systems. IEEE Transactions on Human-Machine Systems, 43, 608–619.

Hollnagel, E., Woods, D.D. & Leveson, N.C. (Eds.) (2006). Resilience engineering: concepts and precepts. Aldershot, UK: Ashgate.

Houghton, R. J., Baber, C., Cowton, M., Walker, G. H. & Stanton, N.A. (2008). WESTT (workload, error, situational awareness, time and teamwork): an analytical prototyping system for command and control. Cognition, Technology and Work, 10, 199–207.

Houghton, R.J., Baber, C., McMaster, R., Stanton, N.A., Salmon, P.M., Stewart, R. & Walker, G.H. (2006). Command and control in emergency services operations: a social network analysis. Ergonomics, 49, 1204–1225.

Harvey, C. & Stanton, N.A. (2014). Safety in System-of-Systems: ten key challenges. Safety Science, 70, 358–366.

Jenkins, D.P., Stanton, N.A., Salmon, P.M. & Walker, G.H. (2010). A systemic approach to accident analysis: a case study of the Stockwell shooting. Ergonomics, 53, 1–17.

Lundberg, J., Rollenhagen, C. & Hollnagel, E. (2010). What you find is not always what you fix – how other aspects than causes of accidents decide recommendations for remedial actions. Accident Analysis and Prevention, 42, 2132–2139.

MAA 2010. RA 1210: Management of operating risk (risk to life). In: Authority, M. A. (ed.) RA 1210. London: Military Aviation Authority.

MAA 2011. Regulatory instruction MAA RI/02/11 (DG) – Air safety: risk management. London: MAA.

Moray, N., Groeger, J. & Stanton, N.A. (2017). Quantitative modelling in cognitive ergonomics: predicting signals passed at danger. Ergonomics, 60(2), 206–220.

Noyes, J. & Stanton, N.A. (1997). Engineering psychology: contribution to system safety. Computing & Control Engineering Journal, 8 (3), 107–112.

Plant, K.L. & Stanton, N.A. (2012). Why did the pilots shut down the wrong engine? Explaining errors in context using Schema Theory and the Perceptual Cycle Model. Safety Science, 50 (2), 300–315.

Rafferty, L., Stanton, N.A. & Walker, G.H. (2012). The human factors of fratricide, Surrey, Ashgate.

Ramos, A.L., Ferreira, J.V. & Barcelo, J. (2012). Model-based systems engineering: an emerging approach for modern systems. IEEE Transactions of Systems, Man and Cybernetics: Part C – Applications and Reviews, 42, 101–111.

Rasmussen, J. (1997). Risk management in a dynamic society: a modelling problem. Safety Science, 27, 183–213.

Read, G.J.M., Salmon, P.M., Lenné, M.G. & Stanton, N.A. (2015). Designing sociotechnical systems with cognitive work analysis: putting theory back into practice. Ergonomics, 58, 822–851.

Royal Navy. 2012. Fleet requirements air direction unit (FRADU) [Online]. Royal Navy. Available: http://www.royalnavy.mod.uk/sitecore/content/home/the-fleet/air-stations/rnas-culdrose/fleet-requirements-air-direction-unit-fradu [Accessed 11.01.2013 2013].

Salmon, P.M., Cornelissen, M. & Trotter, M.J. (2011a). Systems-based accident analysis methods: a comparison of AcciMap, HFACS, and STAMP. Safety Science, 50, 1158–1170.

Salmon, P.M., Stanton, N.A., Lenné, M., Jenkins, D.P., Rafferty, L. & Walker, G.H. (2011b). Human factors methods and accident analysis. Surrey: Ashgate.

Salmon, P.M., Lenné, M.G., Walker, G.H., Stanton, N.A. & Filtness, A. (2014). Using the event analysis of Systemic Teamwork (EAST) to explore conflicts between different road user groups when making right hand turns at urban intersections. Ergonomics, 57, 1628–1642.

Stanton, N.A. (2006). Hierarchical task analysis: developments, applications, and extensions. Applied Ergonomics, 37, 55–79.

Stanton, N.A. (2014). Representing distribution cognition in complex systems: how a submarine returns to periscope depth. Ergonomics, 57, 403–418.

Stanton, N.A., Baber, C. & Harris, D. (2008). Modelling command and control: event analysis of systematic teamwork. Surrey, UK: Ashgate.

Stanton, N.A., Harris, D. & Starr, A. (2016). The future flight deck: modelling dual, single and distributed crewing options. Applied Ergonomics, 53, 331–342.

Stanton, N.A., Rafferty, L.A. & Blane, A. (2012). Human factors analysis of accidents in systems of systems. Journal of Battlefield Technology, 15, 23–30.

Stanton, N.A., Salmon, P., Harris D., Marshall, A., Demagalski, J., Young, M.S., Waldmann, T. & Dekker, S.W.A. (2009). Predicting pilot error: testing a new methodology and a multi-methods and analysts approach. Applied Ergonomics, 40 (3), 464–471.

Stanton, N. A., Salmon, P.M., Walker, G. H., Baber, C. & Jenkins, D.P. (2005). Human factors methods: a practical guide for engineering and design (first edition), Aldershot: Ashgate.

Stanton, N. A., Salmon, P.M., Rafferty, L.A., Walker, G.H., Baber, C. & Jenkins, D.P. (2013). Human factors methods: a practical guide for engineering and design (second edition). Aldershot: Ashgate.

Stanton, N.A. & Stevenage, S. (1998). Learning to predict human error: issues of reliability, validity and acceptability. Ergonomics, 41 (11), 1737–1756.

Stanton, N.A., Stewart, R., Harris, D., Houghton, R. J., Baber, C., McMaster, R., Salmon, P., Hoyle, G., Walker, G.H., Young, M.S., Linsell, M., Dymott, R. & Green, D. (2006). Distributed situation awareness in dynamic systems: theoretical development and application of an ergonomics methodology. Ergonomics, 49, 1288–1311.

Stanton, N. A. & Walker, G.H. (2011). Exploring the psychological factors involved in the Ladbroke Grove rail accident. Accident Analysis & Prevention, 43 (3), 1117–1127.

Stewart, R., Stanton, N.A., Harris, D., Baber, C., Salmon, P., Mock, M., Tatlock, K., Wells, L. & Kay, A. (2008). Distributed situation awareness in an airborne warning and control system: application of novel ergonomics methodology. Cognition, Technology and Work, 10, 221–229.

Underwood, P. & Waterson, P. (2013). System accident analysis: examining the gap between research and practice. Accident Analysis and Prevention, 55, 154–164.

Von Bertalanffy, L. (1950). An outline of general system theory. British Journal for the Philosophy of Science, 1, 134–165.

Walker, G.H., Gibson, H., Stanton, N.A., Baber, C., Salmon, P. & Green, D. (2006). Event analysis of systematic teamwork (EAST): a novel integration of ergonomics methods to analyze C4i activity. Ergonomics, 49, 1345–1369.

Walker, G.H., Stanton, N.A., Baber, C., Wells, L., Gibson, H., Salmon, P.M. & Jenkins, D. (2010a). From ethnography to the EAST method: a tractable approach for representing distributed cognition in air traffic control. Ergonomics, 53, 184–197.

Walker, G.H., Stanton, N.A., Salmon, P. M. & Jenkins, D.P. (2008). A review of sociotechnical systems theory: a classic concept for command and control paradigms. Theoretical Issues in Ergonomics Science, 9, 479–499.

Walker, G.H., Stanton, N.A., Salmon, P.M. & Jenkins, D.P. (2009). Command and control: the sociotechnical perspective. Aldershot, UK: Ashgate.

Walker, G.H., Stanton, N.A., Salmon, P. M., Jenkins, D.P. & Rafferty, L.A. (2010b). Translating concepts of complexity to the field of ergonomics. Ergonomics, 53, 1175–1186.

Waterson, P., Robertson, M.M., Cooke, N.J., Militello, L., Roth, E. & Stanton, N.A. (2015). Defining the methodological challenges and opportunities for an effective science of sociotechnical systems and safety. Ergonomics, 58 (4), 565–599.

Wilson, J.R. (2012). Fundamentals of systems ergonomics. Work, 14, 3861–3868.

Chapter 5

Worksite risk management as a cultural issue

The needs of risk work and regulation to improve safety culture

Eric Drais

INRS, French Research and Safety Institute for the
Prevention of Occupational Accidents and Diseases
Vandœuvre-lès-Nancy, France

CONTENTS

5.1 INTRODUCTION

For the workplace to be safe, operators must be in a constant state of alertness. Although technical systems and work organization are generally designed to reduce potential exposure of workers as much as possible, risks persist, and accidents occur regularly. Incidents are often linked to a lack of alertness and attention among individuals. Indeed, workers' responsibility is often invoked, suggesting a developed capacity for risk regulation, or a greater or lesser degree of intentional risk-taking. Nevertheless, individuals are never isolated from their environment, and alertness and attention are both individual and collective concepts, as demonstrated by sociology and management studies relating to a culture of safety or security.

The notion of culture is expressed at an organizational level through coordinated attitudes and practices about health and safety and has become strategic in the field of industrial or technological risk. These areas of risk

DOI: 10.1201/9781003221609-5

have allowed numerous studies of performance, reliability, or resilience. None of these studies report a single, perfectly shared culture. We thus describe collective parameters, with factors relating to both insecurity and safety. In this field, our own research into prevention culture indicates that health and safety are closely linked to the collective prevention approaches governing the activity within a company. However, it appears that this work of prevention, which gives rise to the culture of prevention is not often given its full place. As sociologists at Institut National de Recherche et de Sécurité (INRS), we work on promoting the value of this work of prevention. Our studies reflect the presence in organizations, at various levels (local, regional, or national) of varied prevention cultures, including in network structures or while working remotely. The underlying work of prevention particularly relies on implementing a strategy for organization and negotiation with respect to the risks, which once again relates to the risk work developed by certain authors (Alaszewski, 2018).

Based on the case of a company that wishes to investigate the alertness and behavior of its personnel to guarantee safety, the objective of this chapter is to present how organizational variables related to prevention can be recorded to act on risk (and alertness) management through cultural management. Cognitive sciences are currently adding value to the notions of alertness and attention, and it appears opportune to us to review their respective roles in safety, and how they are regulated collectively. We will first present the context of the need for alertness, followed by a pedagogical reflection to encourage the company to overcome the simple behavioral question, to allow its managers to consider developing a culture of prevention to preserve workers' health and safety.

5.2 ALERTNESS, ATTENTION, AND RISK-TAKING: ARGUING FOR A SYSTEMIC AND CONTINGENCY-PLANNING APPROACH

Today, risk management in the workplace relies on multidisciplinary and multi-stakeholder approaches. From this point of view, the advantage of sociology is that it provides an understanding of the interactions taking place within the occupational systems. In this context, based on their individual characteristics, alertness, and attention (see definitions in Box 5.1) can be used to perfectly illustrate the collective dimensions of any organized action. These concepts also remind us of the importance of taking the work context into account for health and safety, as well as the social and productive functions of organizational cultures.

Alertness and attention are two essential concepts that are associated with scientific and management considerations of systems presenting risks: any activity involving driving (machinery, installation, process, etc.)

BOX 5.1   ALERTNESS AND ATTENTION: REMINDER OF DEFINITIONS

People use alertness and attention to adapt their behavior to the environment (Broadbent, 1958). They have long been associated to the point that they have been linked to demanding tasks with respect to awareness or concentration (e.g., radar monitoring tasks in aeronautics). Historically, performance in "alertness tasks" for radar operators constituted one of the first scientific fields of research into alertness and attention. These tasks are characterized by their relatively long duration, rare detection needs, difficult-to-predict signals among other stimuli. Quality control in industry constitutes another field of interest for operational research. The most-commonly used means for measuring performance in "alertness" tasks is the proportion of target stimuli detected (e.g., defective products in industrial output). The question of alertness is raised for numerous working activities, where quality and safety are at stake over long periods of time, of medical doctors or surgeons for example.

It is important, however, to distinguish between alertness and attention, which are often confused. In summary, alertness describes a psychophysiological capacity to take in (perceive) information and respond to a situation. It is related to a general state of functional activation for the individual. The history of armed conflicts has led to continuous research for means to optimize alertness in soldiers, more specifically through the administration of synthetic drugs. Beyond these specific circumstances, optimal alertness is also sought in the field of occupational risks, in particular for individuals working atypical hours.

Attention, in contrast, describes the cognitive capacity to verify and handle information. It designates resources and processes that guide and specify this information processing. Classically, three types of attention can be distinguished: selective attention, which is the capacity to select from among the information available; divided attention, which refers to the capacity to share attentional resources across several tasks; and sustained attention, which is the capacity to maintain one's attention over a long period.

Performance and perturbations of the alertness/attention system have been the subject of multiple multidisciplinary research studies. Chronobiology and chrono-psychology first showed that numerous physiological and psychological parameters follow a circadian rhythm (over 24 h), thus establishing a "temporal structure" for the individual (Reinberg, 1974). Assessments of alertness levels in these fields indicated that this capacity reaches its minimum at around 5 a.m., then increases rapidly until the end of the morning, increases more slowly until the end of the afternoon, and subsequently decreases rapidly until nightfall (Thayer, 1978). Apart from this very variable level of alertness, another interesting finding from these studies is that they showed a systematic "drop in alertness" over time, with reduced attention recorded sometimes after just a few minutes' activity.

generally calls on a worker's capacities to remain alert or attentive. Thus, the desire to optimize resources in the field constitutes a common preoccupation for performance in companies, in terms both of productivity and safety, and particularly for systems presenting a risk. Although legitimate, this preoccupation nevertheless has the effect of concentrating on the individual without focusing much on the organization. An opposing conception exists as a foil to this purely psychological or cognitive conception, and we are interested in defining a systemic approach to situations representing a risk in the workplace which reveals the interrelated dimensions of exposure to risks. A culture of prevention, which links back to standards, representations, and values associated with health and safety, is a means of understanding and getting to grips with the dimensions influencing the risks, while also determining the attitude that should be adopted.

The performance of the alertness/attention system relies on personal and organic factors linked to the physical or chemical capacities for information transmission, which are themselves highly interlinked (fatigue, nourishment, stress, addictions, etc.). In addition, the detection of stimuli is also linked to how easy it is to discriminate between signals (their potential to be perceived). Physiological or generic cognitive parameters must therefore be considered in combination with factors specific to the situation (shiftwork, atypical or alternating working hours, etc.) and to the characteristics of the task (monotony, etc.). Orientation with respect to individual strategies and collective learning is also involved in detection. These dimensions can be understood in terms of culture and experience, in that the skills required to identify and detect issues, linked in particular to the work of prevention, can be developed.

Current theory and existing data in this field appear to indicate that it is possible to promote performance by acting on alertness. This approach concerns the improvement or facilitation of perceptive discrimination, choice of appropriate strategic adjustments, prevention of the effects of lack of attention, maintenance of a standby state, etc. Studies relating to "Crew Resource Management" – well known in the aeronautics sector – have borne witness to this possibility for a number of years. Their results show that it is possible to adapt individual alertness levels from the stage of design of systems, by creating organizational conditions of teamwork that will help direct attention. The notions of crossed or shared alertness used in numerous safety models and approaches to safety also illustrate this movement (including for safety of the public, see Roux, 2006). The advantage of these notions is that they establish a collective approach to alertness, allowing individual attention to be directed to promote safety. It therefore appears necessary to us to review organizational cultures of prevention to understand how risks are dealt with in such situations, like in the example below.

5.3 THE CALL FOR ALERTNESS BY DEFAULT

The example that we present stems from a request for advice from a company specializing in maintenance and exploitation of the French road network. The company's board of directors had questions about the performance of its health and safety management system, despite its OHSAS 18001/ISO 45001 accreditation, and wished to increase alertness on its work sites. Indeed, against all expectations given the efforts deployed, serious accidents persisted. Whereas the danger was peripheral to the activity, exposure to it, linked to road traffic had led to regular accidents – including lethal accidents every two years – over the last five years. The directors – who considered that they had pushed their organizational approach to its limits – wished to encourage operators to be more alert to this threat, external to their own tasks. As the main part of the activity performed on these work sites (which are often mobile) is linked to road maintenance, the management considered deploying an intervention device based on the implementation of "overseers" to monitor work sites, who would receive dedicated training.

Among the work situations presenting a risk, four activities drew our attention: an activity to ensure road safety through the placement of safety barriers along the edges of the road; two road maintenance activities, trimming plants on the verge with a "pruning team" (in the form of a convoy of trimming machines, see illustration below), and pruning and chipping waste (Figure 5.1); and finally a marking activity (placement and removal of traffic cones to delimit work sites, often in an exposed context and sometimes reversing against normal traffic-flow). All these situations raise the issue of site mobility, and thus questions related to protection of the site as its activity progressed, with the simultaneous requirement to monitor and manage road traffic (Box 5.2).

In these varied work contexts where configurations depend on the road network, alertness is an absolute requirement, and is assumed to guarantee a high level of safety on sites, through better adaptation of the activity and

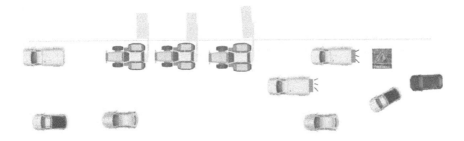

Figure 5.1 Pruning team on highway road.

BOX 5.2   THE CASE OF ROAD WORKS SITES

The company managing the road network employs more than 1000 staff. This personnel includes workers, technicians, and management grouped together not only in intervention, maintenance, and work functions, such as patrollers, study managers, heads of projects, and works inspectors, but also support functions (HR, prevention, etc.). This staff is spread across 28 centers within the geographical territory covered – 4 French regions and 12 departments – to manage 1700 km of national roads and motorways. The company has been interested in the prevention of occupational risks for several years; it is OHSAS 18001/ISO 45001 certified.

At the time of the initial request, the project was to reinforce safety on work sites by specializing some personnel and systematically giving them a specific "overseer" mission or function, to monitor traffic and the road environment. This organization was imagined as a complement to behavioral training as part of an overall safety program to incite the whole intervention team to greater alertness.

The request we received related to the plan for safety management and led us to set up a mentoring relationship with the company over several months. We first asked to get to know the activities so that we would fully understand the need for alertness and the stakes. After a review of the documents relating to risk assessment, the health and safety and accident management system, we met the teams (managers and personnel) involved in the activities presenting a risk. We selected work situations for observation and documentation which were then presented for analysis by participants to produce collective safety rules during training sessions in the context of participatory workshops.

---

constant adjustment to the environment as well as to traffic. The company's management consider it a specific, additional skill, that can be delegated to one person and shared by the group. Indeed, the management believes that the call to alertness should make information on the situation accessible in real time so as to use this information to readily solve problems. We were intrigued by this conception of alertness and safety. As they have implemented a complete system to manage health and safety, including multiple procedures and requirements, it appeared that the management considered the prescriptions sufficient, and thus they relied online-managers and teams to take responsibility for local adaptations. This certainly corresponds to the models that are currently used, and particularly the idea that safety is a combination of regulated safety (resulting from prescriptions developed in advance or set out in regulations, by technology, etc.) and managed safety (developed in the situation by the professionals to meet safety needs locally)

(Morel, 2008). Indeed, in the field of ergonomics, it is common to observe a distinction between the work prescribed and the work performed effectively, and the importance of the latter. For company's workers, the sections of road are often part of a continuously changing environmental context, from the point of view of topography, the weather, or road traffic. However, should we consider it futile to attempt to consider all the working situations and anticipate issues in a general sense? Work should not necessarily be regulated by an incomplete set of rules. It appeared interesting to us to examine situations requiring specific alertness, and to investigate the question of how the rules regulating the work are defined. Indeed, the absence or incompleteness of regulations is a recurrent problem in prevention of occupational risks. This problem differs depending on activity sectors and contexts and can be solved one case at a time only. Designers must therefore constantly examine the activity context to identify the best responses in terms of adjustments to be made for safer production. In any case, the occupational health and safety procedures cannot simply consist in communicating the regulations to the frontline workers and exposing them to risks. In this context, a detailed activity analysis, with the workers, is justified.

5.4 ANALYSIS OF WORK AND RELATED RISKS, A CHALLENGE FOR MANAGERS AND DESIGNERS

With the desire to understanding the real working conditions, we observed work sites and created video recordings of the activity which were then presented to the teams. Classically, we analyze work by examining the task and scrutinizing the activity. More specifically, task analysis focuses on objectives, instructions, and instruments, as well as protocols. In contrast, activity analysis is more broadly interested in the workplace and relations between workers' characteristics (e.g., physical, mental) and activity's sociotechnical conditions related to tools, instruments, social, spatial, temporal organization, and so on.

Examination of the videos rapidly revealed that not only teams from the various canters did not agree on the definition of the tasks to be completed, but also that activity and associated risks were considered differently. For example, site protection was a subject of debate: was it better to work during the day or at night to reduce exposure to traffic? Was it better to totally cut off traffic during work or to reduce the number of traffic lanes? Should sites be protected by simple marking or by setting up physical barriers with machines? Teams then discussed operating modes, current methods, not only defending both distinct ways of working and standard approaches, but also debating the extent of the risks and thus how sites should be prepared. In addition, we observed that participants also welcomed advice or training offers for the analysis of working situations. Although they already had a checklist for site preparation, requiring a site inspection and analysis of the working conditions, they nevertheless expressed the need for additional elements of response to the following types of question: what kind of analysis of the working situation? How is it performed? Using what method? With whom? For what purpose? On what type of activity? And so on.

To complete this activity analysis, we discussed the question of alertness in the context of work and the risks identified. From the accident analysis point of view, at least three contextual questions must be considered: is the activity risky? Can activity-related risk(s) be controlled by people? Is the working situation familiar and routine, or is it new and evolutive?

Indeed, alertness and attention vary depending on the nature of exposure to danger. When the danger is evident, attention is obviously expected. In general, when the main danger is central to the activity, attention is required at all times: this is the case, for example, for mountain climbers, pilots, sailors, lumberjacks, fire personnel, or soldiers in combat. They are faced with obvious dangers that are inherent to their activity, and which must be reduced, mastered, and controlled. In this case, there is no dissociation between safety and production, attention is inherent to the activity performed and to its safety. Occupational behavior is thus closely linked to this primary requirement to ensure the safety of the personnel, but also because the activity's success depends on it.

Attention can nevertheless be put on hold as this type of situation simultaneously exposes workers to the need to develop psychological defense mechanisms or "profession-specific defensive ideologies" (see Dejours, 1980). These mechanisms or ideologies are a means to view the danger in perspective or to place some distance between the worker and danger to reduce fear (potentially due to a perception of permanent exposure to hazards, over which they have little control). Sometimes, there is a tendency to refuse (or even make a joke of) formal safety recommendations as they paradoxically contribute to feeding/re-igniting fear (by "breathing life into danger").

Attention can also be divided when danger – even obvious danger – becomes peripheral to activity. This is the case in the activities mentioned in road works, where safety and production are dissociated (after all, the objective is not to eliminate road traffic!). In this case, the need to elicit alertness to traffic is almost exclusively related to prevention objectives. This is effectively what the company wishes to develop, but it nevertheless indicates that the motivation and resources available for prevention are very different from those available for production. In addition, if attention is divided, the difficulty remains that the constant reminder of the reality of the danger can lead, as above, to the creation of the opposite to the expected attitudes.

Attention is also linked to whether, or not, it is possible to control the dangerous situation. Can the dangerous phenomenon be eliminated or reduced by the individual or through organizational means? Depending on the organizational context and the available technology, how the risk management is distributed between the individual and their work context varies considerably. Thus, situations exist where safety is strongly linked to the operator's behavior (e.g., isolated work in a forest), or the opposite, where working situations involve work or safety considerations that are almost completely independent of the individual's behavior. Attentional requirements will therefore differ in each case. Systems assisting with piloting or navigation in civil aviation must deal with this need for case-relevant adaptation. According to the "integrated safety" concept, the sociotechnical system takes responsibility for safety. Even though the danger is far from eliminated, this offloading of responses progressively dampens the worker's sense of being exposed to danger, leading to a drop-off in attention (through prolonged absence of perception of obviously dangerous situations). In the case of road works sites, the difficulty is also linked to the key problem of uncertainty and variability in critical events over time. Thus, the protective equipment required fluctuates over time, and decisions rely to a greater or lesser extent on the company or individuals.

Finally, the third variable is that of the regularity of the situation in which the work is conducted: is it recurrent and routine or, is it new or dysfunctional? The routines – often observable in repetitive work contexts or when risks become rare (but not necessarily less severe) – require less sustained attention and can rely on simplified decision-making modes (Rasmussen, 1997) or derivatives (errors, mistakes, violations, etc.). These approaches are often justified by human operators as facilitating their activity or eliminating aspects they consider not to be useful (for reasons of physical comfort, waiting times or circuit durations, "added" safety, etc.).

We note that these adjustments, although more or less clandestinely implemented, are often tolerated by managers. Thus, the alteration often constitutes "one of the aspects of appropriation of the task" (Monteau, 1997, p. 37), whereby workers take initiatives, or go beyond prescriptions, with the proviso, however, that the expected production is not compromised (either qualitatively or quantitatively). In addition, when risk-taking corresponds to an attempt to recover from an incident, and this attempt is successful, it is generally accepted that there is no cause for complaint! It nevertheless remains that it is a question of a need to direct attention based on knowledge or perception of potentially dangerous situations.

New situations can also lead workers to take risks by omitting procedures or instructions related to their work if they view them as insufficient or inappropriate. This is the case for example in working situations which do not run according to plan. Unexpected events and other perturbations lead to attempts to recover (Faverge, 1970) which can lead, in turn, to incidents or accidents (when recovery fails). In contrast to routine situations, attention is used selectively, including control and prevention of incidents (Amalberti, 2013). Both situations are, indeed, potential sources of deviations which could become incidents.

For the company being studied, the issue of prevention appears identical for all activities (i.e., based on the need to manage both risks inherent to the activity and road risks, external to the activity). Behind the question of alertness, the management also aimed to increase the attention its teams paid to road traffic, which it considered as a recurrent external variable associated with accidents on work sites. However, as we have just seen, in the situations reviewed, road traffic requires both attention and alertness, as well as interaction with the other variables associated with the activity.

5.5 SAFETY AND PREVENTION CULTURE: AN ORGANIZATIONAL CONTINGENCY

Although attention can vary and there exist numerous psychological or cognitive mechanisms which can affect decision-making, it is nonetheless important to create the conditions to direct attention toward the task, the group, and the situation. This process is mainly a question of agreeing on potential risks and alerting workers to allow their identification and reduction. In this context, it appeared very useful to develop an organizational culture of prevention. This approach is already used to deal with major risks, where it can be seen to develop safety and to regulate attention around the question of occupational risks in general.

Organizational cultures often exist in companies. In terms of safety, as elsewhere, it is often less a question of company culture in response to the wishes or directives of the management, but rather occupational cultures inherent to structures (Douglas, 1982), to the activities performed or the technology used (Liu, 1981). Within these cultures, an experience of risk appears essential. Indeed, in sociology, historical studies indicate that it is by experiencing the risk that it is possible to learn (Sainsaulieu, 1977).

Learning about the risks is about continuously adding professional experience, in concert with more or less formal mentored training initiatives. This is where the work of prevention intervenes. This work, defined by sociologists as a collective activity producing rules to promote/preserve health and safety (Mias et al., 2013), relies on organization (Tersac, 2011), negotiation (Courtet and Gollac, 2020), and adjustment (Strauss, 1978). In part, it covers the notion of risk work that describes how individuals in their daily activities select and give meaning to questions of risks, and then take them into account in their actions (Alaszewski, 2018). The work of prevention makes it possible to develop cultures of prevention. This reliance explains why it is important to verify that this work develops workers' capacity to identify and regulate risks in specific situations. Developing a prevention culture requires not only reviewing the way health and safety is organized, material, measures, and equipment, but also surveying individual and collective risk experiences, both inside the company and with respect to external parameters. A difficulty in performing this review, of course, relates to the availability of or the possibility to create a network, and in what conditions this team can meet up to discuss various aspects. From a cultural point of view, the challenge is to share representations, values, and standards of relationships, with a view to developing coordination. In this context, successful collective preventive work can be used to identify shared sensitivities and lists of collective actions supporting attention and alertness.

For the company presented here, to meet these requirements, we have started a collaborative study relating to accident prevention on sites representing a risk. In contrast to the initial request from the company, we chose not to question, train, and warn the personnel of the need for alertness as such, as we believe it is not enough to ask workers to work safely to make the work safer! It was therefore important to relate the questions of alertness and attention to the context of the activity and the team's work.

On this point, we note that activity is associated with a prevention culture, not only from a micro-sociological point of view, but also at a more macro-sociological scale at the level of safety systems. At the international scale of industrial or professional sectors and their organizations, it is important to remember that three main approaches to safety are presented in the literature for systems representing a risk (Amalberti, 2013). These approaches reveal three attitudes or almost archetypal strategies, with respect to safety or security: the first consists in eliminating or systematically avoiding exposure to the danger. This is the case, for example, with "ultra-safe" systems (more specifically transportation such as the aviation industry) where, aware of the risks, the slightest danger is banished as a principle through the application of rules and procedures (postponement of a flight or alteration of the flight path in case of climate events, etc.). The second consists in tolerating exposure to danger on the condition that the rules and procedures instituted are followed. This is the case of systems representing major industrial risks (e.g., nuclear, chemical) where complex processes involved require means of control relying on so-called high-reliability organizations (or HRO). The third and final attitude is that of accepting not only risks, but also specific work of adjustments made by operators when faced with a risk. This is the case with activities such as deep-sea fishing where a strict application of safety recommendations and measures could, in principle, prevent any activity. The activity in turn is closely linked to system's (boat, team, etc.) capacity for resilience. These contexts and safety models will necessarily dictate the proportion of unregulated risks that persist for the activity, and where designers and managers can intervene in complement to the workers' own decision-making process. The extensive development of regulations governing means of production, and the associated safety requirements must thus be considered and maintained from the point of view of these distinct attitudes to safety (Bourrier, 2018) and the contingency of the various activity-related organizational variables.

5.6 CONCLUSION

Since Descartes, the attention phenomenon has occupied a place between sensation and desire, raising questions more specifically related to the models used and safety conditions: "We are both masters and slaves of our attention. We can direct and focus it, but it can also escape us, be captured by external events or objects" (Lachaux, 2011). With a view to creating safe systems, complementary to prevention by design, it is important to investigate cultures of prevention by examining the conditions in which prevention is implemented and the collective capacity to identify, raise awareness of, and regulate occupational risks. The challenges related to implementing a culture of safety in a company are to develop, through coordinated organizational actions at various levels, a reflection that will raise questions in the company related to prevention at every level of the organization. The needs associated with developing a health and safety culture in a company are to guide and assist the development of representations to allow safe and healthy activity for all those working in the company. In practical terms, for the company, this requires work on the presentation of the elements that will elicit, mobilize, and grab the attention of workers to preserve their health and safety, or that of the company. While favoring organizational monitoring of safety elements with the possibility to implement safety management systems, the company will attempt to clearly identify and communicate on elements. In the day-to-day management of occupational risks to which workers are exposed, these elements may be controlled at the organizational level of the work (regulated safety) or left to individual or collective management decisions (managed safety). In this context, the specific place of training and analysis of the work, as transitional spaces for debate between shareholders on occupational risks, must be discussed.

ACKNOWLEDGMENT

This chapter is the result of an action-research project conducted in collaboration with Marc FAVARO (Psychologist, PhD, Senior Research Scientist, now retired, INRS). We would like to thank him for his knowledge and his inexhaustible contributions in the field of prevention, used in particular in this book. We remember his long and stimulating multidisciplinary discussions. We also thank Jacques MARC (Psychologist, PhD, Senior Research Scientist, INRS) for his reading.

REFERENCES

Alaszewski, A. (2018). Tom Horlick-Jones and Risk Work. *Health, Risk & Society*, 20(1–2), 13–22.
Amalberti, R. (2013). *Navigating safety: Necessary compromises and trade-offs-theory and practice* (Vol. 132). Heidelberg: Springer.

Bourrier, M. (2018). Safety culture and models: "regime change". In: *Safety cultures, safety models* (pp. 105–119). Cham: Springer.

Broadbent, D.E. (1958). *Perception and communication.* Oxford: Oxford University Press.

Courtet, C., & Gollac, M. (Eds.) (2020). *Risques du travail, la santé négociée.* Paris, France: La Découverte.

Dejours, C. (1980). *Travail: usure mentale, Bayard, 3èmeéd, 2000.* Paris, France: Bayard.

Douglas, M., & Wildawsky, A. (1982). *Risk and culture: An essay on the selection of technological and environmental dangers.* Berkeley, CA: University of California Press.

Faverge, J.M. (1970). L'homme, agent de fiabilité et d'infiabilité du processus industriel. *Ergonomics*, 13, 301–327.

Lachaux, J.-P. (2011). *Le Cerveau attentif. Contrôle, maîtrise et lâcher prise.* Paris, France: Odile Jacob, 368 p.

Liu, M. (1981). Technologie, organisation du travail et comportement des salariés. *Revue française de sociologie*, 22(2), 205–221.

Mias, A., Legrand, E., Carricaburu, D., Féliu, F. & Jamet L. (2013). Le travail de prévention. *Les relations professionnelles face aux risques cancérogènes*, Toulouse, France, Octarès, 200 p. https://doi.org/10.4000/sdt.2487

Monteau, M. (1997). *Prise de risque," dérives" et autres imprudences.* [Rapport de recherche] Notes scientifiques et techniques de l'INRS, NS 155, Institut National de Recherche et de Sécurité (INRS), 161 p.

Morel, G., Amalberti, R., & Chauvin, C. (2008). Articulating the Differences between Safety and Resilience: The Decision-Making Process of Professional Sea-Fishing Skippers. *Human Factors*, 50(1), 1–16.

Rasmussen, J. (1997). Risk Management in a Dynamic Society A Modelling Problem. *Safety Science*, 27, 183–213.

Reinberg, A. (1974). *Des rythmes biologiques à la chronobiologie.* Paris, France: Gauthier-Villars.

Roux, J. (2006). *Etre vigilant – L'opérativité discrète de la société du risque.* Saint-Etienne: Presses Universitaires.

Sainsaulieu, R. (1977). *L'identité au travail.* 2ème édition 1985. Paris: Presses de la Fondation Nationale des Sciences Politiques.

Strauss, A. (1978). *Negotiations: Varieties, contexts, processes, and social order.* San Francisco, CA: Jossey-Bass.

Thayer, R.E. (1978). Toward a Psychological Theory of Multidimensional Activation (Arousal). *Motivation and Emotion*, 2, 1–34.

Chapter 6

Risk-taking: An opportunity to win in elite sports

Anne-Claire Macquet

French Institute of Sports, Laboratory of Sport,
Expertise and Performance
Paris, France

Antoine Macquet

Seine Grands Lacs, Local Administration in Charge of
Flood Prevention
Paris, France

Liliane Pellegrin

French Military Epidemiology and Public Health Center (CESPA)
Aix - Marseille University, IRD, AP-HM, SSA
Marseille, France

CONTENTS

Risk is unavoidable and the inability to cope with risks, and their subsequent opportunities, may prevent athletes from wining. Acceptance of these risks and risk-taking are a fundamental part of decision-making. They depend on situation assessment and the possible consequences of risk, fatigue, emotions, self-confidence, and trust, as well as motor skills. Berg (2010) developed a model of risk management that is well suited to explaining risk management and risk-taking in sports. He stressed four ways of coping with risk: (a) avoid risk, (b) reduce risk, (c) transfer (share) risk, and (d) retain (accept) risk.

DOI: 10.1201/9781003221609-6

Looking at different sports, this chapter examines the ways national coaches briefed their athletes prior to competitions to develop strategies and tactics for adapting to risks during competitions. It also illustrates a process used to prepare for world championships (WCs), using Berg's crisis management model. It highlights the ways athletes assessed potential risks, took risks during competitions, explains the roles of self-(over)confidence and trust, efficacy beliefs, emotions, fatigue, and personality in relation to risk management and risk-taking. At the applied level, perspectives are suggested, including communication, mental, and physical preparedness, for successfully transforming risk-taking in competitions into opportunities to win.

6.1 INTRODUCTION

In elite sports, athletes make decisions in dynamic and complex situations in order to win. Dynamic situations include intense time pressure, ill-structured problems, uncertain environments, competitive goals, decision loops, high stakes, the number of other players, and organizational goals and norms (Macquet, 2021). The path to victory involves risks. The inability to manage these risks represents a danger. When making and implementing decisions, elite athletes expose themselves to risk in order to achieve specific goals (e.g., score, take the advantage in a rally). Risk is the possibility that a potential hazard may materialize (Leplat, 2003). Exposure to the possibility that a set goal either will or will not be achieved creates risk. Risk is therefore unavoidable; it forms an inherent part of situations and actions.

In dynamic situations, athletes must assess the situation rapidly in order to make and implement a decision before it is too late (e.g., before the shuttle falls on the badminton court; Macquet, 2009). Situation assessment can reduce the uncertainty surrounding events. In many dual and collective sports, athletes try to deceive and surprise their opponents by creating uncertainty about their own actions to gain the advantage and score. They feign specific plays to mislead their opponents and limit the time the opponents must assess the actual situation and decide. Such ruses represent a risk for opponents, who must attempt to discriminate between real and fake actions and respond to real actions in time. Athletes thus have two possibilities. They can wait for a situation to develop to assess it with maximum knowledge of the situation and take the risk of making a decision too late. Or they can anticipate the way the situation may develop and assess it rapidly, with the risk that the action taken may prove to be inappropriate. Uncertainty about the outcome of the action may also generate doubt that may block or delay the commitment to a course of action (Lipshitz & Strauss, 1997). In other words, the athlete may ask him/herself: should I stay, or should I go? In dynamic situations, a moment of hesitation by an athlete might enable an opponent to take advantage of the indecision and win.

In competitive situations, athletes make a series of decisions that must be implemented at specific times. Situations evolve at their own pace as well as in relation to the consequences of previous decisions (Brehmer, 1992). Such developments lead to changes in the problems to be solved (e.g., losing the ball causes basketball players to reorganize themselves to recover it). The dynamic development of situations involves risks for athletes. Acceptance of these risks is a key part of decision-making.

In dynamic situations, risks can be internal and/or external (Amalberti, 1996). Internal risks refer to: (a) the risk of not having sufficient skills to achieve the goal (e.g., in a WC, implementing a skill for the first time when it has only just been learnt in training) and (b) the risk of being unable to manage one's own available resources while making and implementing a decision, leading to a loss of control of the situation development (e.g., due to fatigue, a volleyball player may be too late to spot that the ball is being deflected by the block and may not move forward to defend rapidly enough). External risks consist of the risks athletes face as the situation develops (e.g., the rain may make the road slippery and increase the possibility of falling for cyclists) and the risks athletes take when making a decision (e.g., attack the opponent's fencer and take the risk that the opponent may parry-riposte and win the point). Consequently, athletes must find a balance between risk-taking (internal risk in relation to bounded rationality and external risk in relation to target performance) and the consequences (control of a situation and maintaining physical and psychological security, Amalberti, 1996).

Risk-taking is a fundamental part of decision-making. It entails abduction, enabling someone to predict a possible outcome and achieve it by acting appropriately (cf. Introduction). In elite sports, this abduction involves assessment of expertise and motor skills. The level of these skills fluctuates over the competition depending on fatigue and emotions (Macquet, 2021). When they take risks, athletes must take this skill fluctuation into account to achieve their goal.

Risk can impact on a decision in two ways: (a) they relate to situation uncertainty and influence the decision to be made or (b) they relate to their impact on the effectiveness of the decision. In the first case, athletes must manage the risks of situation development and make decisions. In the second case, they take the risk of failing. Both types of risk are embedded in the decision-making process. In elite sports, decision-making thus involves both risk-taking and risk management.

In high-level sports, risk-taking is considered at two different times: before a competition takes place and during a competition. Before a competition, coaches and athletes prepare to adapt to risky situations and potentially take acceptable risks. During a competition, they manage and take acceptable risks to achieve positive outcomes.

This chapter aims to provide an overview of how elite athletes take and manage risks to maximize performance. More specifically, it aims to explain (a) how athletes perceive and assess risks; (b) how they prepare to take and manage risks prior to a competition; and (c) how they take and manage risks during a competition.

6.2 PERCEPTION AND EVALUATION OF RISK

Risk is unavoidable. It results from the effect of uncertainty on objectives (NF ISO 31000, 2018). Exposure to the possibility that a goal either will or will not be achieved creates risk. Risk refers to the possibility that an event, situation or activity may have unwanted and negative consequences upon achievement of a goal (Leplat, 2003). In elite sports, negative consequences denote poor performance, contributing to a decrease in athlete and country rankings and reputation, a subsequent decrease in funding to prepare for competitions, injuries, and so on. Individuals and teams set goals and strategies to achieve goals. They then assess the risks inherent in these strategies to decide whether and how to implement them. The decision-maker either accepts or rejects the risk that he/she might lose control of events, actions, or situations. Risk is a factor the decision-maker takes into consideration: the risk of losing control determines his/her behavior (Oppe, 1988). There are two elements that require the decision-maker's attention: first, risk identification, which aims to find and recognize risks that may jeopardize the achievement of goals (NF ISO 31000, 2018); second, risk management, which enables decisions to be made.

The decision-maker weighs up the expected benefits and possible negative consequences of risk to decide whether to accept the potential risks (Fischoff, 2013). Risk can be acceptable in one situation but not in another if the associated benefits are different. The decision-maker can also take the risk of failing in some parts of a match, assuming temporary failures and disadvantages (i.e., opponents scoring) in order to achieve their main goal later (i.e., winning the match). Risk acceptance involves assessing the balance between the expected benefits of specific actions and strategies and possible negative consequences (i.e., costs). It is about weighing up the advantages and disadvantages, being aware that the decision-maker(s) might or might not achieve the goal, and assessing any circumstances that could jeopardize achievement of the goal. Risk acceptance is thus a two-step process: gauging possible risk and assessing the nature of potential damages if the risk comes to materialize (Cadet & Kouabenan, 2005). The decision-maker assesses which elements of the situation and consequences are acceptable, i.e., tolerable, manageable, or not, and beneficial or harmful, and defines the level of risk acceptance. Risk acceptance makes risk-taking an opportunity to achieve – or not achieve – the goal.

Individual psychological factors including emotions, beliefs, and mental models influence risk acceptance and risk-taking. They enable individuals and teams to be aware of risks and to either over or underestimate or correctly estimate the risks. They may also provide biases in the perception and assessment of risks. Factors such as the illusion of control (i.e., exaggerated perception of the ability to control events), unrealistic optimism (i.e., tendency to more frequently perceive the occurrence of positive events than negative ones), and the illusion of invulnerability (i.e., tendency to perceive oneself as unlikely to suffer the harmful consequences of a danger) tend to result in the underestimation of risk and lead the decision-maker to neglect or be blind to risk. Such effects were shown in a study of professional athletes (e.g., Karatas, 2016).

Moreover, organizational, economic, political, and cultural factors also influence risk acceptance (Wiese-Bjornstal, 2019). Diversity in terms of beliefs and mental models due to the level of expertise, roles (i.e., leader, follower), and personality (e.g., optimistic vs. pessimistic) may lead to differences in situation assessment, risk acceptance and risk-taking among athletes, and strategies and actions to be implemented (Martha & Laurendeau, 2010). Beliefs and mental models may also affect individual's perception of the credibility of preventive measures and allow individuals to trust such measures and have confidence. Such preventive measures are generally designed by experts, managers, authorities, and individuals who assess acceptable risk and manage risk (Kouabenan, 2009).

Risk management enables decisions to be made with the aim of succeeding at individual or collective level (Berg, 2010). Within an organism or specific structure, such as a sports team, it is carried out at group level. Organisms or structures thus have a determining role in risk management. Berg (2010) identifies seven steps in the risk management process: (a) establish goals and constraints (e.g., qualify for the Olympic Games); (b) identify the risks (e.g., fail to qualify in a specific Olympic qualification tournament); (c) analyze the risks (e.g., what needs to be done to qualify at different stages? What are the risks at each stage?); (d) evaluate the risks (acceptable risk criteria, e.g., if the team has not qualified at this time and there is still another way to qualify, risk may consequently be tolerable); (e) deal with the risk (e.g., how can the team increase the chances of qualifying?); (f) monitor the risk (e.g., check whether the strategies and tactics are being adapted and implemented effectively); and (g) communicate and report on risk (e.g., debrief on strategies and tactics used during the match and players' activity and attitudes). Berg (2010) highlighted four ways of dealing with risk in organizations: (a) avoid risk, (b) reduce risk, (c) transfer (share) risk, (d) retain (accept) risk.

Such risk management outside the domain of sports is well suited to explaining risk management in sports. It can be used as a lens through which to view studies explaining decision-making in sports in situations where athletes face risks. In high-level sport, risk is managed on two occasions: prior to a competition and during a competition.

6.3 PREPARATION FOR TAKING AND MANAGING RISK PRIOR TO A COMPETITION

Prior to a competition, coaches of team and dual sports brief their athletes to develop strategies and tactics to manage risk, cope with time pressure and uncertainty, and adapt to the opponent's game (Lainé, Mouchet, &

Sarremejane, 2016; Macquet & Stanton, 2021). To prepare for a match, coaches and athletes set goals consisting of high-performance expectancies (i.e., win the match). They use video sequences of previous matches and competitions to analyze the opponent's game and assess the opponent's strengths, weaknesses, opportunities, and threats (SWOT; Lainé et al., 2016; Macquet & Stanton, 2021). Furthermore, they identify the opponent's action (individual sports) and pattern of coordination (team sports), to help them design a successful strategy. In doing so, they identify risk in different areas of the game using mental simulation. They also analyze risk by investigating what makes an opponent's action risky (e.g., has the volleyball defender enough time to move toward this zone to defend the ball?), and evaluate the risk by asking whether it is acceptable or not (e.g., did the opponent often hit that zone in previous matches?).

Coaches and athletes deal with risk by investigating the various ways in which the athletes could adapt their play to suit the situation. They may (a) avoid the risk by moving the block to protect a specific zone (and take the risk of making another zone more vulnerable); (b) reduce the risk by asking a defender to pay more attention to a specific attacker's hitter's opportunity for action than to other players and to be ready to adapt if the opponent hits differently; (c) transfer the risk by changing defender in order to position a more effective defender in that zone; and (d) accept the risk that the opponent's team may score by hitting that way, for example, if the opponent seldom hits toward this area, the risk may be negligible and have only a small impact on the overall score.

Risk is assessed during briefings including staff and athletes (Macquet & Stanton, 2021). A frequent scenario involves each participant watching a video on his/her own; then he/she shares his/her analysis during meetings of the group. He/she anticipates possible situation developments in light of the risk. For example, in team sports, coaches and athletes plan game actions and assess the risks on performance, and then set out the chosen actions in a game plan. The game plan is an adaptation of the playbook defining plays (i.e., structured patterns of teammates' coordination and actions) and players' roles in relation to their competencies within the team (Eccles & Johnson, 2009). Each player knows what he/she must do, and his/her actions can be predicted by his/her teammates. Plays enable (a) players' actions to be organized to deal with opponents' risky actions and (b) risk to be limited by deciding on actions that best use the players' skills. In deciding on a specific play, coaches and players consider the risk that the opponent could adapt and score. However, plays are flexible, and players adapt them to the situation development, deciding whether to continue or change a play depending on perceived risk (Macquet & Kragba, 2015). Coaches mentally stimulate players by asking "what if?" questions (e.g., if the ball comes from that side, what will you do? Macquet & Stanton, 2021).

To adapt to risk, players and coaches plan tactics they are used to implementing. Such tactics refer to well-practiced and effective procedures used in routine-centered and structured situations. Sometimes these tactics may be ineffective in risky situations, leading coaches and players to think creatively to solve a potential problem. Creativity is developed during training sessions (Memmers, 2015) and briefing in order that athletes can adapt to nonroutine situations. Coaches accordingly encourage athletes to analyze opponents' SWOT, and consider what they may be able to do in order to develop the game plan. Such analysis and a problem-solving approach help athletes to face and manage risk. Macquet and Stanton (2021) showed that coaches empowered athletes to make their own analysis and be responsible for the game plan and risk-taking. Together, they devised new ways of coping with risk. They also watched video sequences involving different opponents in similar situations, as well as analyzing how opponents adapted in previous matches and then assessing whether their own athletes could adapt in the same way. They used video sequences to gain inspiration from the experience of other athletes and decide whether to adopt the opponents' behaviors or create new angles. Such collective problem-solving includes brainstorming. Brainstorming is a key element in preparing individuals to cope with uncertainty and risk. Collective problem-solving is conducted by the head coach, and involves assistant coaches and athletes in a meeting. In team sports, it is conducted with the whole team, as well as with smaller groups formed according to players' roles (i.e., forwards in football, setters in volleyball). Such brainstorming enables players to be empowered to take risks they can manage both physically and mentally. Mental simulation is then used to assess whether the decision is workable and appropriate. Brainstorming is followed by a short training session with the aim of implementing what was decided and ensuring that the athletes can take and manage risks.

The aim of briefing is to develop tactics designed to fulfill expectations of the outcome by taking appropriate action. The development of tactics considers external risks linked to the situation development prior to any decision. It also considers internal risks, such as the risk of not having adequate skills to achieve the goal, and the risk of mismanaging one's own resources while making and implementing a decision. Such risks might lead to a loss of control of the situation development. Coaches and athletes gauge risks to make good decisions and take risks in carrying them out.

Macquet and Stanton (2021) showed that coaches took athletes' needs and capabilities when coping with risks into account in briefing and inspired confidence. Confidence plays a key role in helping to ensure a plan can work and that the athletes can adapt it (Vealey & Vernau, 2010). Coaches talk about similar situations where athletes took a risk and achieved positive outcomes. They use teambuilding and communication to foster team cohesion and build athletes' confidence (Vealey & Vernau, 2010). They inspire athletes and support them so that the athletes can commit themselves in a particular situation and take risks in order to achieve positive outcomes. Coaches and lead athletes behave like leaders, guiding their players and teammates, empowering them to solve problems, and being closely involved with the team to ensure the work is achieved, while at the same time looking after the well-being of the team. Exploring the effects of confidence on risk-taking could be of interest for future research.

Macquet and Stanton (2021) highlighted that intellectual stimulation and collective problem-solving enable athletes to cope collectively with possible risk-induced stress. Adopting Leprince, D'Arripe-Longueville, and Doron's (2018) work on communal coping strategies (e.g., problem-focused communal efforts, relationship-focused coping, and communal management of emotion), we could expect that athletes collectively identify and share stressors and make efforts to cope with stressors by solving the problems raised or by withdrawing collectively. Such strategies enable athletes to cope with potential stress while facing risk.

Most frequently, staff and athletes use strategies based on routines and experience to achieve their goals. They adapt strategies to cope with problems arising during action. In some cases, the initial strategies prove to be ineffective and must be changed significantly. The 2004 French female canoe sprint team provides a case study for such changes. In 2004, the team was 26th in the world rankings. In 2005, the staff and athletes started to manage and take risks in relation to the WCs in Zagreb. They conducted a process in line with Berg's (2010) crisis management process.

First, they set a main goal consisting of giving their very best right up to the finish line, in spite of any pain caused by high intensity muscular effort. They left performance goals that had been defined in previous years and that consisted of an expected specific ranking at the end of the race. They rather set mastery goals based on commitment. In giving their very best, staff hoped that athletes might perform better and limit any performance anxiety caused by performance goals. They also identified the constraints in preparing for the WC and achieving this goal. Such constraints related to (a) the short length of the preparation period; (b) climate, limiting the training possibilities; (c) available funding; (d) the politics of the federation and the French Ministry of Sport; and (e) the opponents' quality of preparation.

Second, staff and athletes identified the possible risks for athletes in the WC. Positive consequences included maximum commitment by athletes, enabling them to improve their international ranking, self-confidence and trust in staff and athletes, and the preparation strategy for the WC, as well as bonuses through winning medals and enhanced status (i.e., being on the elite athlete list of the French Ministry of Sport, which provides advantages in terms of training and quality of life). Negative consequences included opposition and possible defiance toward the staff and future strategies that might detract from commitment and performance.

Third, staff and athletes analyzed the risks to decide whether they were acceptable. They developed a consistent strategy led by a manager who was well regarded within the group. They all discussed the strategies and risks to ensure that everyone felt part of the team. They organized regular meetings to answer questions related to potential or actual problems and find solutions together. They divided the work to be done and responsibilities within the team and decided on specific actions for everyone. For example, coaches were responsible for contacting the referees if there was a dispute involving athletes during the race, such as a boat hitting and being slowed down by a water lily. Coaches were also responsible for asking referees to arrange for the waterway to be cleaned before the next events. Athletes were responsible for giving their very best while paddling, despite unexpected events and pain.

Fourth, staff and athletes used three ways to manage the risks (a) acceptance of risks; (b) mitigation of risks; and (c) transferring risks. They were aware of the risks and did not try to avoid them rather they accepted them. To mitigate risks, coaches and each athlete set a race strategy, depending on the athlete's abilities, mental and physical preparedness, and state of mind. They briefed and debriefed every day about what went well and what did not, possible causes, and how to maintain or improve individual and collective activity. Their aim was for athletes to improve steadily through the WC. Athletes transferred part of the risks to the staff and the preparation strategy, as well as to their teammates. They collectively decided how to divide the responsibilities and work in relation to their roles. Together they were responsible for overall performance, while they transferred responsibilities and risks for specific operations to individuals (e.g., contacting referees was the responsibility of coaches, giving their very best was the duty of athletes).

Fifth, they frequently debriefed on the risks and strategies used. They debriefed both after and between WC events in order to (a) take into account what had happened, the risks encountered and how they had coped with them and (b) set optimal individual and team race strategies for the following events and improve race performance. For example, during the debriefing of a WC series in 2005, the female team reported that their boat had got caught on a water lily in the middle of their race. The plant had slowed their boat down. The athletes said that they had persevered despite the resistance of their boat and expended much greater effort to finish the race, battling through the pain. The coach congratulated them for their mental toughness. Such mental toughness enabled them to qualify for the following events and eventually, to win the bronze medal. Staff and athletes continued to implement these strategies during the following WC and noticed that this change of preparation strategies for the WC, combined with the empowerment of athletes and coaches, enabled them to improve the international ranking of France, going from 26th place in 2004 to 6th in 2010. Thinking outside the box and taking risks enabled them to improve their performance and ranking over the years.

6.4 TAKING AND MANAGING RISK
DURING A COMPETITION

During a competition, when making a decision, athletes assess the risk associated with the situation they are involved in and the risks inherent in any decision they take. Macquet and Kragba (2015) showed that basketball players assessed possible risk to decide whether or not they should undertake a play. In the beginning of a game situation, the playmaker plans and calls for a play by showing his/her teammates a sign with his/her fingers. Each player knows what he/she must do. Plays are flexible and players are required to adapt them to the situation development and decide whether to continue or change them depending on the perceived risk. Macquet and Kragba (2015) showed that players first assessed the current situation development by checking whether the current situation was developing as expected in the playbook. They also simulated the possible consequences of their planned decision on the situation development to determine if risk was absent, manageable, or difficult to manage. When no risk was perceived, they continued the play. When risk was considered to be manageable, they continued the play, but were more involved when facing the opponent's action (i.e., they reduced the risk). When risk was assessed as difficult to manage, players changed the play to avoid taking the risk that may have prevented the play (i.e., they avoided the risk). Mental simulation was used by the players to assess risk for the team.

In the same vein, in time pressure situations, team sport players were shown to use mental simulation in order to assess whether their decision was workable (i.e., potentially successful with low risk of failure, e.g., Kermarrec & Bossard, 2014; Macquet, 2009). If they considered it was too risky, they avoided the risk by making another decision. Results highlighted that team sport players seldom reported anticipating risks related to performance. This suggests that time pressure was so high that players did not have time to investigate the risks linked to several options before choosing one. Exploring the impact of risk perception on decision-making in high time pressure situations could be a worthwhile avenue for future research.

According to Berg (2010), athletes treat risks in the course of action in action in four ways. First, they avoid risk when risk is not tolerable, meaning when they assess that they have no chance or little prospect of achieving a successful outcome. In doing so, they change the play or tactics. Second, athletes reduce risk by using a specific tactic and by being more involved than usual to achieve a positive outcome. For example, a fencer pays a lot of attention to the position of his/her opponent's trunk and tries to put a lot of pressure on his/her opponent to push him/her back. Third, athletes share and transfer risk by avoiding implementing an action. For example, in badminton, the front player crouches down to let his/her teammate defend the shuttle. Fourth, athletes may accept risk and prepare to lose control of the situation. For example, a volleyball player may assess a ball as going out of the court and not defend it, taking the risk that the ball is actually in the court and that the opposing team might score. Such ways of dealing with risk are practiced in training sessions. Players learn to read the shuttle or ball trajectories in changing contexts involving high uncertainty and time pressure, as well as fatigue. Fencers use individual sessions with their coach to learn how to deal with a specific risk. The coach replicates specific tactics that pose a risk for his/her athlete and forces his/her athlete to adapt. He/she reproduces the situation and then changes the tactics for the athlete to learn to adapt effectively. After each exercise, the coach provides feedback on the athlete's action, relevant cues for the situation and/or asks the athlete to recall what he/she perceived, thought, did, and so on. Individual training is used to practice managing risk and risk-taking in implementing newly learnt actions. These sessions provide an opportunity for brainstorming about actions; they enable actions to be implemented and feedback to be provided to improve both action and performance. Athletes thus learn to adapt to very specific situations regarding both action and communication.

To manage and take risks, athletes use well-practiced tactics. Tactics refers to procedures defining what to do, when, where, how, with whom, and for how long. They consist of motor and mental skills enabling athletes to make swift decisions and succeed in known situations. Athletes learn to be flexible while using their skills to adapt to dynamic situations. Skills provide a way to reduce risk-taking in familiar situations. When a situation is unfamiliar, athletes must change their tactics and may not have appropriate tactics to adapt. In that case, they must either create something new or implement an existing skill in a new way. Creativity enables them to vary the game (Memmers, 2015). This need for creativity means athletes must think outside the box. For example, in Macquet and Hermet (2013), a world champion orienteer explained the strategies he used to read a map and navigate in uncertain environments without being seen by opponents. He developed procedures to simplify the map to navigate and use the shortest and most efficient running routes. When discrepancies appeared between the map and the environment, his existing procedures did not enable him to navigate efficiently and rapidly. So he took the risk of pausing during the race to stand back from the map and seek relevant cues in the environment. Such risk-taking enabled him to obtain an overall sense of the route and continue to navigate safely.

Athletes implement specific procedures to achieve goals in familiar situations. They are also required to deviate from procedures when they feel the procedures will not enable them to perform well. In that case, they must create new ways to adapt and act. Adaptation is a strategy used to protect ourselves against risk (Klein, 2009). Choosing between known procedures and innovation is fraught with risk for athletes. It poses even more risk in team sports than individual sports because members of a team must be able to predict their teammates' actions. To ensure such predictability, coaches provide players with roles. They allow one specific player to change part of the playbook in the course of action. Coaches say that they provide a specific player with a "gold card" permitting them to rearrange the playbook. Rearranging the playbook goes beyond adapting the play as described above; the playmaker may create new plays and take initiatives in the course of action. In order to limit risk-taking and be predictable by teammates, only one player is provided with this role.

The 1996 French canoe slalom team provides a useful example of initiative in the course of action. Before the final of the Atlanta Olympic Games, Franck Adisson and Wilfrid Forgues had prepared the course with their staff. While paddling during the race, the French athletes were faced with a roller, where the main current reversed its flow upstream. In their race plan, they had planned to attack the roller from the front. However, during the descent, their boat drifted. They could have continued their planned trajectory and risked wasting time and missing the next gate. However, the rear paddler noticed the disadvantageous position of the boat on the river and paddled fiercely to make the boat move backwards and recover its position in order to continue on its course. The front paddler sensed the change in the boat's movement; he adapted by synchronizing his paddling with his teammate's. They did not communicate as they did not have enough time, they just felt the boat move, which was a relevant cue for both. The rear paddler thought creatively for a way to regain their position and adapt the boat's trajectory to the unforeseen situation. This was an exceptional maneuver that they had never implemented before, but taking this risk enabled them to become Olympic champions.

Four years later, in Sydney, they aimed to be crowned Olympic champions for a second time. However, this time, the Olympic rules had changed; the race times of the semi-final and final were combined. In the semi-final, the paddlers picked up a two-second penalty, placing them behind their main opponents. They had to gain time to win. Taking tighter trajectories would enable them to gain time. At one of the last gates, they decided to depart from their race plan and tighten their trajectory, instead of using a looser route. It meant that they might gain on their opponents' time and win the race (i.e., opportunity) but at the same, it brought the risk that they might touch or miss the gate and get another penalty, meaning walking away from the gold medal (i.e., disadvantage). They took the risk and gave it everything they could. This change in their course resulted in the back paddler touching the gate and getting a two-second penalty. In the end, they lost rather than gained time and came seventh at the OG. If they had stuck to the plan, they would have won silver. This risk-taking caused them to lose their Olympic title and prevented them from winning a medal. Risk-taking brings opportunities to win. Such opportunities attract athletes, who also accept the corresponding possibility that they may lose.

As previously mentioned, dynamic situations evolve on their own as well as in relation to the consequences of previous decisions (Brehmer, 1992).

Such development includes changes in the problem to be solved and may jeopardize achievement of the goal. In sport, changes are also connected to the uncertainty athletes create to deceive their opponents. Athletes make feints to create uncertainty and provide less time for their opponents to rapidly make an appropriate decision. They pretend to make a play and then change tack to trick their opponents into making a mistake. For example, a volleyball player jumps to hit a rapid ball at the center of the net; his/her teammate immediately jumps behind him/her to hit a "slow" ball. If the opponents jump to block the first hitter, they are coming back down when the second hitter changes tack and attacks; consequently, they fail to block the second ball. Athletes try to be as credible as possible in these feints, pretending to play a certain way and then changing tack to trick their opponents into making a mistake. In practice, athletes learn to spot feints and not respond. Faking improves with experience; however experts also learn to discriminate between what is real and what is fake, taking the risk of failing by choosing the wrong option and/or implementing an effective action too late. Some athletes are very skilled at faking, which makes the use of feints high chance of deceiving the opposition and thus the feint succeeding.

In nonroutine situations, such as feints, experts must be flexible to adapt to uncertainties, detect anomalies, reframe the situation (Klein, Philipps, Rall, & Peluso, 2007), and distinguish between relevant and irrelevant information. They use affordances to assess the situation. Perception in such situations is guided by mental models, consisting of schemata (Neisser, 1976). Schemata determine what the individual is capable of perceiving and whether information refers to activated schemata. Perception also follows a cyclical pattern. The result of the interaction between the individual and the environment changes primary schemata that, in turn, changes perception. When the opponent makes a feint, the athlete seeks to detect anomalies to spot the feint. Athletes try to know when the opponent makes a feint; at the same time, the opponents try to make feints realistic. Athletes learn to compromise between understanding the situation rapidly, which triggers an automatic response, and slower understanding, which allows anomalies such as feints to be detected. In the first case, they take the risk of being wrong in responding to a fake they failed to identify as such. In the second case, they take the risk of spending too much time focusing on a potential fake and not having time to carry out an appropriate action. Training and briefing enable athletes to notice when a feint is being made, and how to adapt. Making feints and noticing feints both involve risk-taking, which in turn involves availability, attention, and open-mindedness at all times.

Risk-taking is also a matter of self-confidence and trust. To take risks, the athlete must be confident in his/her own ability to achieve a goal. Low self-confidence and trust, as well as low self-esteem and low self-efficacy belief comprise internal risks. According to Amalberti (1996), internal risk relates to being unable to manage one's own resources while making and implementing decisions, leading to a loss of control of the development of the situation. Self-confidence fluctuates throughout a match for athletes, depending on their performance, with the strongest source of confidence being success (Veley & Vernau, 2010). Confidence is derived from mastering and improving personal skills, demonstrating one's abilities to others, physical and mental preparedness, social support from others, vicarious experience, and a supportive environment. Self-confidence is built with the help of others (i.e., sport staff, teammates, family, friends) and the environment.

Moreover, efficacy beliefs enable individuals to believe they can exercise control over a situation; this can boost their confidence (Bandura, 1997). In practice, athletes and coaches report that athletes take more risk when they feel confident and trust their teammates and coach. Conversely, they take less risk when they are less confident; they then favor their own proven tactics, making them more predictable to their opponents. Trust and self-confidence are important in sport and more specifically toward the end of a match when the score is tied. In such a situation, taking risks can enable an athlete to win when risks are well managed or to fail when they are not. However, athletes need to ensure that they are realistic in assessing situations. Optimistic bias seems to play a role in risk-taking as many athletes who misread the risk overestimate the chances of success rather than failure (Kahneman, 2011). Such overconfidence might act as a barrier to appropriate decisions and adaptation. Exploring the effects of self-confidence and self-efficacy belief could be of interest for future research on risk-taking in sports and other contexts.

In the same vein, anxiety dissuades athletes from taking and managing risk. Anxiety involves a decrease in attention and a certain degree of blindness due to hypovigilance. Anxiety and emotions in general influence perception and decision-making (Mosier & Fischer, 2009). Emotions affect risk-taking too, leading individuals to either take or avoid risk (Bonnet & Pedinielli, 2011). Managing emotions is thus an important part of decision-making and risk-taking. Research on the effects of emotions on decision-making could be a fruitful avenue for future research (Macquet, 2021).

Fatigue is also known to influence risk-taking, leading athletes to take either more or less risk because the workload is too high and as a result, they are in a state of hypovigilance. To combat this, training aims to delay the onset of fatigue. In competition, emotions and high-performance expectations also influence fatigue and workload. Athletes thus need to keep their workload to a reasonable level by focusing on what is relevant and ignoring what is not.

Finally, from a psychological perspective, questions have been raised about the relationship between personality and risk-taking (Bonnet & Pedinielli, 2011). It is thought that sensation-seeking, impulsivity, or aggression could help explain an individual's tendency to take risks. Risks can be denied, ignored, or chosen (Bonnet & Pedinielli, 2011). Risks may be denied or ignored when the individuals are overconfident. Chosen risk brings a challenge: the person selecting it is asserting their belief in their ability to control a situation. The individual might have the illusion of control over a situation.

Self-confidence, emotions, and fatigue accordingly all have an effect on risk-taking. They can be considered either strengths or vulnerabilities. It is therefore important for athletes to know when and how such factors influence their decision-making and risk-taking. Enabling individuals to know themselves better is important in high-level sports and other contexts. Mental preparedness is a useful resource in this area as it can help people to understand themselves better and improve mental skills such as self-confidence, concentration, stress management, and, more broadly, regulation of emotions.

At the applied level, and looking beyond mental preparedness, it is important to develop research into communication and reporting about risk-taking. Clear communication is fundamental in the risk management process (Berg, 2010) and risk-taking. It relates to required actions and feedback about results. Reporting is another element in the process. Any reporting requirements should be set out in debriefs with coaches and athletes and in documented procedures (Macquet, Ferrand, & Stanton, 2015). The content of such communication may include an analysis of the taskwork, including identified risks, what was done, and what went well and badly, to understand what happened and why and set athletes' requirements for the following competition.

6.5 CONCLUSION

Life without risk does not exist. Nor does opportunity without risk. As stated in NF ISO 31000, risk-taking can be considered both an opportunity and a threat depending on the anticipated outcomes. In high-level sports, athletes and have little chance of winning if they take a zero-risk approach. The zero-risk approach means not trying. Athletes must weigh up their ability to implement any decision in a specific situation and the costs and benefits of the consequences of the decision. In doing so, they gauge tolerable risk and make the decision either to take the risk and deal with it, or to ignore it. Unfortunately, ignoring it does not necessarily mean the athlete can avoid it. Risks must be managed or coped with.

Athletes and coaches manage risk twice: before a competition and during a competition. Before a competition, they assess possible risks, and plan how to change and be flexible in order to adapt to unexpected risks. Collective brainstorming enables risks to be assessed and athletes to prepare to adapt in order to implement risk tactics and face risky situations. During a competition, athletes manage and take risks under conditions of intense time pressure and uncertainty, using their skills and self-confidence to control situations and hopefully win. When the time pressure is too high, they do not have time to assess the risk. The more athletes have prepared for risk-taking and risk management, the more likely they are to reach their goals. To be able to take risks and achieve positive outcomes, athletes and coaches must rely on training, skills, mental and physical preparedness, briefing, and a pinch of luck. Developing risk management and risk-taking procedures with individuals and teams is a major challenge, but one that can create winning opportunities.

As such, risk management in elite sports is not limited to competition. Training, both mental and physical, allows coaches and athletes to perfect their skills to successfully transform risks taken in competitions into opportunities. In the same vein, risks relating to, for example, injury, exhaustion, or non-sports-related events or trauma affecting the athlete's performance may occur during training or even in their personal lives. Studying risk-taking in other contexts would thus be a worthwhile avenue for research and training programs.

ACKNOWLEDGMENTS

The authors would like to thank Christophe Rouffet and Bertrand Daille for their useful comments on their experiences in canoeing.

REFERENCES

Amalberti, R. (1996). *La conduite des systèmes à risques* [Managing risky systems]. Paris: PUF.

Bandura, A. (1997). *Self-efficacy*. New York, NY: Freeman and Company.

Berg, H. P. (2010). Risk management: Procedures, methods and experiences. *RT & A*, 2(17), 79–95.

Bonnet, A., & Pedinielli, J.-L. (2011). Conduites à risque et prises de risques en contexte sportif [Risk conduct and risk-taking in sports contexts]. In G. Decamps (Ed.) *Psychologie du sport et de la santé* (pp. 331–347). Bruxelles: De Boeck.

Brehmer, B. (1992). Dynamic decision-making: Human control of complex systems. *Acta Psychologica*, 81, 211–241.

Cadet, B., & Kouabenan, D. R. (2005). Évaluer et modéliser les risques: apports et limites de différents paradigmes dans le diagnostic de sécurité. *Le Travail Humain*, 68(1), 7.

Eccles, D. W., & Johnson, M. B. (2009). Letting the social and cognitive merge. New concepts for an understanding of group functioning in sport. In S. D. Mellalieu, & S. Hanton (Eds.) *Advances in applied sport psychology* (pp. 281–316). London: Taylor & Francis Group.

Fischhoff, B. (2013). *Risk analysis and human behavior*. New York, NY: Routledge.

Kahneman, D. (2011). *Thinking fast and slow*. New York: Farrar, Strauss and Giroux.

Karatas, O. (2016). A research into evaluation of basketball athletes' risk perception level. *International Education Studies*, 9(5), 108–114.

Kermarrec, G., & Bossard, C. (2014). Defensive soccer players' decision-making: A naturalistic study. *Cognitive Engineering and Decision-Making*, 8(2), 187–199. doi: 10.1177/1555343414527968

Klein, G. A. (2009). *Streetlights and shadows: Searching for the keys to adaptive decision-making*. Cambridge: Massachusetts Institute of Technology.

Klein, G. A., Philipps, J. K., Rall, R. L., & Peluso, D. A. (2007). A data frame theory of sense-making. In R. R. Hoffman (Ed.) *Expertise out of context* (pp. 113–155). New York, NY: Lawrence Erlbaum Associates.

Kouabenan, D. R. (2009). Role of beliefs in accident and risk analysis and prevention. *Safety Science*, 47(6), 767–776.

Lainé, M., Mouchet, A., & Sarremejane, P. (2016). Le discours d'avant-match des entraîneurs de rugby: des temps d'intervention enchâssés [Before match speech from rugby coaches: shortened intervention times]. In B. Lenzen, D. Dering, B. Poussin, H. Denervaud, & A. Cordoba (Eds.) *Temps, temporalité et intervention en EPS et en sport* (pp. 199–219). Bern: Peter Lang.

Leplat, J. (2003). Questions autour de la notion de risque [Questions around the notion of risk]. In D. R. Kouabenan & M. Dubois (Eds.) *Les risques professionnels: évolutions des approches, nouvelles perspectives* (pp. 37–52). Toulouse: Octarès.

Leprince, C., D'Arripe-longueville, F., & Doron, J. (2018). Coping in teams: Exploring athletes' communal coping strategies to deal with shared stressors. *Frontiers in Psychology*. doi: 10.3389/fpsyg.2018.01908

Lipshitz, R., & Strauss, O. (1997). Coping with uncertainty: A naturalistic decision-making analysis. *Organizational Behavior and Human Decision Processes*, 69, 149–163.

Macquet, A.-C. (2009). Recognition within the decision-making process: A case study of expert volleyball players. *Journal of Applied Sport Psychology*, 21, 64–79. doi: 10.1080/10413200802575759.

Macquet, A.-C. (2021). Decision-making in sport: Looking at and beyond the Recognition-Primed Decision model. In P. Salmon, S. McLean, C. Dallat, N. Mansfield, C. Solomon, & A. Hulme (Eds.) *Human factors in sports and outdoor recreation* (pp. 134–154). Boca Raton, FL: CRC Press.

Macquet, A.-C., & Hermet, A. (2013). Quand les "choses se passent bien" – Quand des difficultés apparaissent... La prise de décision en sport de haut niveau [When things go well, when difficulties appear... Decision-making in high level sport]. In C. Van De Leemput, C. Chauvin, & C. Hellemans (Eds.) *Activités humaines, technologies et bien-être* (pp. 349–354). Bruxelles: Presses Universitaires de Bruxelles.

Macquet, A.-C., & Kragba, K. (2015). What makes basketball players continue with the planned play or change it? A case study of the relationships between sense-making and decision-making. *Cognition, Technology and Work*, 17(3), 345–353. doi: 10.1007/s10111-015-0332-4.

Macquet, A.-C., & Stanton, N. A. (2021). How do head coaches brief their athletes? Exploring transformational leadership behaviors in elite team sports. *Human Factors and Ergonomics in Manufacturing and Services Industries*, 31(5), 506–515. doi: 10.1002/hfm.20899.

Macquet, A.-C., Ferrand, C., & Stanton, N. A. (2015). Divide and rule: A qualitative analysis of the debriefing process in elite team sports. *Applied Ergonomics*, 51, 30–38. doi: 10.1016/japergo.2015.04005.

Martha, C., & Laurendeau, J. (2010). Are perceived comparative risks realistic among high-risk sports participants? *International Journal of Sport and Exercise Psychology*, 8(2), 129–146.

Memmers, D. (2015). *Teaching tactical creativity in sport.* New York, NY: Routledge.

Mosier, K. L., & Fischer, U. T. (2009). Does affect matter in naturalistic decision-making? In Proceeding of the NDM9. London.

Neisser, U. (1976). *Cognition and reality: Principles and implications of cognitive psychology.* San Francisco, CA: Freeman.

NF ISO 31000 (2018). *Management du risque* [Risk management]. www.afnor. org.

Oppe, S. (1988). The concept of risk: A decision theoretic approach. *Ergonomics*, 31(4), 435–440.

Veley, R. S., & Vernau, D. (2010). Confidence. In S. J. Hanrahan & M. B. Andersen (Eds.) *Routledge handbook of applied sport psychology* (pp. 518–527). Oxon: Routledge.

Wiese-Bjornstal, D. M. (2019). Psychological predictors and consequences of injuries in sport settings. In M. A. Anshel & T. Petrie (Eds.) *APA handbook of sport and exercise psychology.* Washington, DC: American Psychological Association.

Chapter 7

Risk-taking: Submarine experience

Ludovic Loine
SAS AXION
La Roque d'Anthéron, France

CONTENTS

7.1 INTRODUCTION

I started my 16-year career in the French Navy after one year at the Military Petty Officer School in Brest as a basic electrician on a Sub-Surface Ballistic Nuclear (SSBN) submarine. The SSBN is the most powerful and complex submarine ever. After several missions at sea, I entered the Military Nuclear School in Cherbourg where I graduated as a Nuclear Specialist operating a nuclear engine on Sub Surface Nuclear (SSN) submarine in Toulon, as Chief Engineer of Nuclear Engine at sea.

Being a submariner is a commitment. It is commitment in a risky professional life. In theory, it is easy to be a submariner with respect to risk being taken. There are only two conditions to be hired as a submariner: being a volunteer and medically able to do the job. With a romantic vision, when you are not familiar with the subject, submarine looks like the Nautilus as the vision of Harper Goff in the movie Twenty Thousand Leagues Under the Sea (a modern adaptation of Jules Verne's novel). As a matter of fact, many submarines throughout the world are named "Nautilus".

If Jules Verne's submarine is a "classic" submarine regarding propulsion, SSBN and SSN are propelled by a nuclear reactor. SSNs have been

DOI: 10.1201/9781003221609-7

developed because of it can operate with remarkable discretion and provide powerful attack possibilities.

SSBN missions are totally different than SSN ones. SSBNs are used for strategic missions, they carry a very specific payload: Ballistic Nuclear Missiles that can strike multiple targets anywhere in the world. An SSN can only carry conventional weapons.

SSN missions are hunter-killer missions, targeting enemy submarines or war ships. Such missions were run during Cold War between East and West blocks. After the fall of Berlin's Wall, missions on SSN changed a little bit and new kinds of threats raised-up, such as Islamic Radicalism or conflicts consecutive to the dismantling of East Block.

I participated into such missions.

Even though submarine crewmembers are everyday confronted to risk, they are not movie stars, and unfortunately, submarine designers did not follow Jules Verne's requirements. Windows and portholes are cruelly missing to fish-, shells- and octopus-watching amateurs, which could have been interesting in helping crewmembers to support long mission out of home. Enough with the jokes.

Modern submarines have more automation than the previous generation. Submarine crewmembers are decreasing in number while submarines are becoming bigger. It is now more comfortable to live in, but it also is riskier due to no permanent human zone. The development of new sensors was necessary, and the complexity of the newer systems increased.

As a submariner, I can tell you that risk-taking is not at all related to the common popular vision, even if most people say that they could not do this job (Boy & Brachet, 2010; Loine, 2014a).

Adding to the difficult context of lack of privacy for long periods, distance with family six months a year, hostile environment (i.e. fighting position) and regarding to the variety of situations (i.e. navigation under sea, or in low deep water), managing the complexity of systems is an everyday task.

7.2 BEING A SUBMARINER

This section will introduce you to how to become a submariner. Engineering designers should understand submariner's culture and background when they are committed to develop a new system.

Why does a sailor decide to become a submariner? The first reason is two-fold: interest for the job and sense of homeland service. Each sailor onboard is responsible of complex systems and is accountable, with respect to other sailors, for the success of the mission regardless of his role. Highly specialized on their job and very well trained to execute it, submariners can face responsibilities, sometimes at a very young age. They might have to bear the life of the whole crew into their hands depending on their position and the situation. Each crewmember has a role onboard and is fully qualified for it. I think being young is very positive for this job; young people are open to take dynamic necessary decisions and to obey on orders. Obviously, they are physically stronger which helps to resist to a stressful environment during a long period of time.

In addition, all submarine crewmembers need to get experience on managing risk on board. They usually start their career at the lower hierarchical levels and make progress in the same environment throughout their entire career, with companionship of older sailors. They improve their knowledge step by step in a strong spirit framework.

Training starts at the Specific Submarine School where each sailor must qualify. The first period at sea with specific training completes the cursus, with the help of older submariners. Risk management skills improve during each mission at sea year after year.

Crew members are rather young with respect to their responsibilities. The average French submarine crew is between 26 and 30 years old.

The usual maxim used is "Prepare for the worst and live for the best". The atmosphere on board is most of the time positive and no hierarchical signs are visible, even though everybody knows the chain of command. As a professional requirement, goal-driven rules are the way to follow. It is essential to trust each other during normal operations as well as uncommon situations. Relationships onboard are genuine and wholesome. When tensions appear between people, the group solves the problem. Each sailor is not alone and can find support from others. It is true both for personal and professional problems. Psychology of the crew is indeed crucial. Psychology of each individual and group dynamics results from both individuals and the current situation. Everybody on board focuses his attention on psychology. I was not aware of that when starting, but it proved to be real after several years of practice.

The organization is the same on all submarines, except for the fact that crewmembers may change jobs from one mission to another. In this environment, responsibilities are given priority over individuals. Even if each crewmember is replaceable in terms of skills, everyone takes care about everyone. The organization is very strong and resilient thanks to that rule.

There are three different teams onboard that are in charge of the submarine on a 24/7 basis. Team building is a major requirement that considers various kinds of experience, sensibility and skills, personality, and people behavior need to be considered onboard.

The rhythm of work onboard is related to the position of the submarine. Daily tasks are shared (i.e., premises cleaning, maintenance of systems, and repairing). Sharing the daily tasks contributes to maintain a friendly atmosphere and generates the feeling to be part of the group.

7.3 SUBMARINER RISKS

What are the risks onboard? As it is the case for space trips, most important risks are related to the complexity of the environment and the vessel. A submarine includes many complex systems. All systems are necessary to enable the crew to do its mission with maximum security, safety, and efficiency. Risks onboard are literally everywhere since the submarine navigates in a 3D environment, likely to an aircraft or a spacecraft. As a submarine is totally isolated from outside facilities, it needs systems that enable it to work independently from any support for a long period of time. Systems are designed to be easily supervised, still some situations make submariners requiring help from outside. We can easily relate to Apollo 13 mission, when engineers had to provide the space crew a solution to build a system that could settle the critical carbon dioxide issue in two hours with a minimum of material. When the *Kursk* Russian submarine sank in 2000, after two big explosions occurred on the front of the ship, there were still crewmembers alive at the rear waiting for rescue. No one had the chance to be rescued in time, despite dedicated rescue systems (i.e., against carbon dioxide) and external means of rescue (i.e., former D.S.R.V).

Let's have a look at the risks that submariners might encounter. We can distinguish:

- risks supported by vessel integrity against external or internal hazards (e.g., weapons, fire, subsea mountain collision, and ship collision). History can provide many examples of submarines bumping onto subsea mountains. It was the case with USS San Francisco (SSN-711), which collided with a subsea mountain of 560 meters high (Figure 7.1). The event was about 350 miles South of Guam (2005/01/08, 02.00 AM)

Figure 7.1 Front of the submarine after collision. (Photo: USS San Francisco (SSN-711) – Wikipedia.)

- internal risks possibly caused by system malfunctions (e.g., steam, electricity, pressure, temperature, rate of oxygen or rate of polluting gas such as carbon monoxide and dioxide, freon, and radio-activity in Nuclear Submarine); In 1968, the French Navy lost the Minerve Classic submarine due to flawed design (a heritage from World War 2) – (see Figure 7.2).
- internal risks due to possible bacteria intoxication (e.g., intoxication by food or water, and various disease).

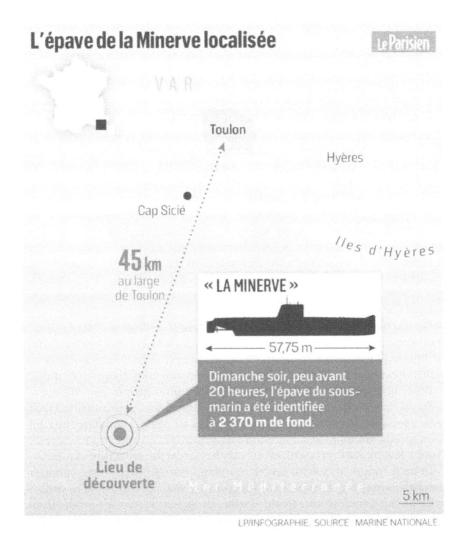

Figure 7.2 Minerve Classic submarine loss. (From www.leparisien.fr)

Atmosphere onboard is continuously controlled with respect to the last case. Air is recycled, and toxic products are controlled because consequences of such aggressions are catastrophic to the crew. From my experience, a few simple events can cause strong dangerous effects. I once had to face one of these events. Food onboard is stored on specific cold room cooled by a dedicated system operating with Freon gas. A leakage of Freon gas is very dangerous for the crew because Freon show hazardous properties when exposed to a heat source. The gas then becomes heavily toxic. When a leakage occurs, it is necessary to confine the compartment and replace the atmosphere using specific system. Systems are often adapted to the situation that had to be managed but not really designed for. When it comes to security issues, as the famous adage says, submariners must "adapt or disappear". In such situation, we changed procedures to adapt and to find an appropriate solution that didn't exist at this time.

7.4 GETTING AND MAINTAINING A SAFETY CULTURE: PROCEDURES

The best way to manage risks and accept to operate with them is to be aware of them, and to trust the systems built to handle the situation. To overcome hazardous events, the crew needs to be very well-prepared: a short amount of time (if no time at all) is available to analyze the situation and respond appropriately. It often is a collective reaction, and each one must know what to do. Procedures exist for each scenario, though it is required to adapt them to the context in real time when an unpredicted event occurs. It is not only about technical or technological reactivity. Organization and leadership models are regularly used for training onboard. For example, our safety culture, specifics skills, missions' experience are constantly considered for improvement.

In practice, appropriate level of risk-taking is defined as "acceptable", with respect to French Ergonomists standards, which considers flexibility or *"room for maneuver"* designed by designers, engineers, managers, as well as health and safety specialists. This approach talks about "soft skills".

Procedures are created with respect to operations people.

By the way, each crew member has a well-defined role, such as being a technical specialist, fireman, first-aid worker, and air analyst. A submariner follows specific training on submersion safety to acquire a specific certification. After a few months on board, everybody has a real safety culture in mind.

In each technical control center of operations on board, three submariners who are specialized on each function are sharing their time to control systems (i.e., nuclear and steam engine, propeller, electricity production). The three of them must be informed and aware of the overall situation, its cause and evolution, to make the most appropriate decision if a procedure came out to fail.

Submariners acquire all submarine safety-related procedures by heart because they need to react very quickly. Nuclear engine procedures are too much complex to be acquired by rote learning, fortunately it is not required to act as promptly. The design of procedures is related to the level of automation, the goal to reach or the event to settle.

Onboard procedures are goal-driven or event-driven. They are regularly improved by integrating experience feedback. Different layers of procedures enable cross-control and limit human errors. The organization allows also to mitigate human errors. The organizational structure is pyramidal with a good definition of what is the perimeter of each crewmember within the organization.

"Safety climate" is another important factor for the robustness of the organization. I would define a safety climate as good working atmosphere as well as enthusiastic and professional team spirit on board. Getting the best safety climate demands perfect respect of subordinates and officers to act suitably on unpredictable events, to avoid "cliff effects". It is a result of strong team building. Sharing the same work rules, the same experience, getting used to work as a team and offer the support of a great organization at each level are necessary to obtain safety climate. In fact, it is not only limited to submarine operations, but also properly the history of collective rules building (local team and professional community).

7.5 CRISIS MANAGEMENT

However, knowing very well risks in a submarine is never enough. The new generation of submarines integrates Information Technologies. These technologies are not totally mature and may cause cultural gaps. They bring new risks because distance between human operators and systems is bigger than ever before. Submariners must understand more complex safety-critical systems and their driving processes to be able to understand and manage critical events. In specific cases, it can be more important to understand how the safety-critical system itself works than understanding the process (i.e., how is the detection of the risk built?). Adaptation to Information Technologies revolution is necessary because new Information Technologies change submariner's culture. Fortunately, submariners have a short career and crew turnover is three years maximum. Mixing experienced submariners with newcomers enables sharing culture and preparing transition periods.

The more difficult thing when managing an abnormal event is to maintain the logic continuity of service until recovering the normal situation. Many parameters must be considered. The graph presented in Figure 7.3 represents how a crisis occurs.

A simple event (yellow zone) could evolve in orange zone or at least in red zone if the situation is not enough understood or if it is poorly managed or not very well managed in a context evolution and so on.

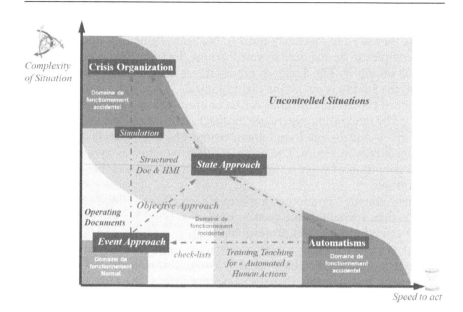

Figure 7.3 Situation complexity vs. action speed. (From Nuclear Engineering Experience.)

Each well-known event is fully documented to be handled properly, avoiding evolutions most of the time in orange or red zones (Figure 7.3). Besides, the organization is adapted with respect to the current context to prepare another future possible evolution of the situation. But everything cannot be totally forecasted. The frame hereunder enables operators and organizations to manage uncertainty with the minimum probability of reaching the red zone as it was the case in Fukushima for example.

A nuclear submarine gathers risks of two complex systems: a submarine and a Nuclear Power Plant (NPP). The best way to make the best decision in an abnormal situation generating discordant objectives relatively to both submarine safety and nuclear reactor safety is to follow only one main objective (MO).

"It is more important for the crew to have a nuclear accident with the submarine at the surface of the water than a reactor shut down with the entire crew dead in deep sea" (Maxim usually promoted on board which is participating to safety climate). In fact, it's a question of life or death. I understand it could be shocking for few people, but it explains why and how the design of submarine is done.

Consequently, there is an impact on nuclear submarine design and the underlying human-centered design (HCD) process.

NPP HCD methods are based on safety-related functional analysis (FA) and task analysis before looking for HMI (Human Machine Interface) solutions. Three major accidents – Three Mile Island (TMI), Chernobyl, and Fukushima – showed that even if people could be the weak link for safety, they are also the last and only resources when everything else has failed. They are also the only resources for crisis management and have avoided most incidents becoming accidents. Most NPP design solutions were implemented more than 40 years ago. New NPP design solutions, such as EPR (Evolutionary Power Reactor), very much improve nuclear safety by considering most of the lessons learnt from major nuclear accident experiences. Today, and very probably in the future, current NPP must face obsolescence and old technologies and, therefore, must be replaced. It is particularly the case for HSI as they are more and more computerized. It is required within safety standards in a continuing process of improvement of safety by implementing new safety requirements and treat aging of life critical systems. In the past, no NPP had been built with respect to HCD principles. What we can presently see in 2020 is that whatever the HMI design, human operators must deal with what was not automated at design time or what engineers forgot. Up to now, most Information Technologies usage was not included in design because of bad experience feedback or no experience at all, because new technology is never mature enough in the first place. Today, nuclear industry must be cost effective in a strong competitive industrial energy world. Nuclear safety is the first issue for both NPP and nuclear submarine for their consequences on populations. Good and successful experiences help for that.

Nevertheless, a submarine nuclear engine more and more follows the same rationale. A submarine is now designed to be safe independently from the nuclear propulsion which wasn't the case in the past. The assessment of the duality between Safety and Availability is no more an issue. Safety 'won the game' because designers integrated this in new nuclear programs by laws. France is the only one country that integrates in the law the precaution principle.

7.6 A FRAMEWORK FOR SYSTEMS DESIGN AND MODELING

Operating a nuclear submarine is a complex collective work. Complexity is present in:

- necessary knowledge regarding human operators' activities, present in the daily work organization (shift work, co-activities to manage);
- global objectives to manage (performance, safety issues, aging, war operations and so on); and
- unpredictable events with important media impact to manage.

Two main contexts need to be considered: War context and Peace context. Starting from this, a large variety of scenarios could be considered for each context. Besides, the framework can frequently and quickly evolve due to changing weather at sea.

It is necessary to rethink the approach of NPP operations to restructure and reorganize activities between human operators on board with respect to all these frameworks. Human operators are not operating only a complex process as fission or steam and electricity production, but they are also operating complex instrumentation and control (I&C) systems not really directly related with their nuclear base knowledge. They are far from the process and shall manage more I&C system than the process itself. It results that is to be sure that human operators understand what is done by I&C and how they should act efficiently on the process.

It is necessary to structure studies during engineering phase in such a way that shall be independent from the technology and propose an overview of functions. That approach of design is a resilient approach that can consider the analysis of cognitive functions and emergent functions or shared cognition. We need a strong dynamics group who increase the competencies of each other and do a collective decision to resolve a problem. The link between individual functions of people and social functions to increase the skills of people is crucial. Individual functions are to be considered as the link between body and mind.

Operators in Nuclear Submarine are not "push button" operators even if more and more procedures would like to automate human actions. As an example, on former French SSN, the whole operating procedures could be sorted in three sorters. Nowadays, it is necessary to have a 2 meters cabinet for sorters. The ongoing evolution of procedures is in my opinion not the best way to take, though it seems it is the way of history. I do not agree with that strategy of organization applied on submarines.

Human operators must analyze and understand what they must do. Procedures cannot always be adapted regarding each situation. That's why the framework shown Figure 7.1 must be implemented during the design of systems. In the case of unpredictable event, appropriate procedures do not really exist. In the case of critical incident or accidental situation, operators operate in a state approach mode coming from lessons learned of TMI. The best way to maintain the real situation awareness in any case is to have a goal-driven approach of operation and procedures to match with the context in real time. Currently, in most NPP over the world – and it is particularly the case in France – this kind of approach of operation is not really implemented. It is the case in Nuclear Submarine. In normal situation or small incident, operators are operating in an event-driven way. However, we can see that between the two approaches, there is often a rupture. Such rupture provokes misunderstanding of the situation and place biases in the whole operating logic which could badly increase the situation instead of improving it.

The concept proposed in that research must cover the whole operational scope in a wide variety of situations. Figure 7.3 in the previous section puts different operations approaches in a diagram addressing situation complexity and action speed. Figure 7.3 summarizes what the human operator does, supported by automatisms or organizational setups. The diagram presents a few means making concrete various possible supports.

The more complex the situation is, the more time is needed by operators to act or organize an action strategy. In most cases, urgency of the response of an event depends on the level of preparedness of human operators, organizational setups, and ability of procedures to match with the current context. If the reaction time is not compatible with human reaction time, automation is necessary. Besides, it assumes that operators can understand automation, robustness, availability, safety classification, and "*Military shock*" (bomb resistance) classification to stay aware of the situation trends and to assess the efficiency of means currently available, and eventually to keep confidence on them.

Consequently, the concept of "Conducting by Objectives" has been developed and implemented using well-adapted digital HMI supporting views of the process and related electronic procedures. This concept helps engineers to manage the complexity of what they must design.

7.7 CONCEPT OF CONDUCTING BY OBJECTIVES

The concept of Conducting by Objectives is based on Cognitive Function Analysis (Boy, 2008). Focusing on the design process, the scope of Conducting by Objectives concept deals with the entire scope of activities in the submarine (nuclear engine operations, navigation, maintenance, lifecycle performance shape factor) and outside the NPP site in case of an accident – see Figure 7.3. This concept can be extended to several layers of the decision power. FA should then be performed considering several functions depending on the frame of operations and the context with respect to the scope of responsibility and accountability of each actor on the decision chain.

For example, let's describe the implementation of the concept in the NPP framework that can be defined by the following two main activities:

- Conducting activity of normal and abnormal operations to achieve current and/or safety objectives.[1]
- Maintenance activity, as an integrating part of NPP activities that can be done while electricity is being produced (it is the case for small maintenance). Big maintenance is done during specific period (outage) when there is no constraint with electricity production. Nevertheless, safety stays as an objective framework.
- It is very important to address simultaneously conducting and maintenance objectives and activities to avoid that human maintenance errors impact conducting activities or, even worst, safety activities.

Management of both activities within the framework of safety rules is led by the definition of key performance indicators (KPIs) to achieve NPP's performance goals. As an example, KPIs would be the number of days in electricity production during a year, the cost of outages, aging monitoring of main components, and so on. Both kinds of activities should be articulated in engineering, and the multidisciplinary team should understand this articulation concept.

Whatever the functions, they are all built on the same way. A conducting function (CoF), a maintenance function (MF) or a safety function (SF) is built around a main physical parameter (PP) or a cohesive group of small systems and components. Each main PP is piloted with means, such as pumps or valves or other components to achieve the function. Of course, secondary PPs are also necessary to monitor components and be able to inform human operators on function availability and operability. Functions are the basis of Conducting by Objectives concept. CoF, MF or SF gives an overall picture of each component allowing managing functions (Figure 7.4).

Functions (conducting, maintenance, safety) are prioritized in a well-known context to adapt conducting views. All functions are targeted to reach the main goals. In the conducting activity, we talk about main objectives as it is presented on Figure 7.5 (Presented at AHFE 2014, Poland).

Main objectives are defined to support NPP operations in context (i.e., electricity production mode, hot shutdown, cold shutdown, and outage). Each main objective that needs to be achieved is supported by COFs. Main maintenance objectives are also defined and are based on a similar concept with MF. A tag allows identifying on which COF a MF could have an impact. specific tools are used to put tags on MF.

There are two priority levels. Priority Level 1 means that the COF is very important to achieve the MO of operation. Priority Level 2 means

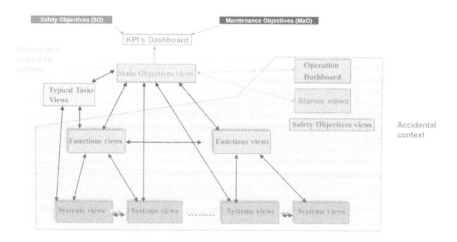

Figure 7.4 Breakdown structure. (From RES [Réacteur d'Essais] Project Implementation.)

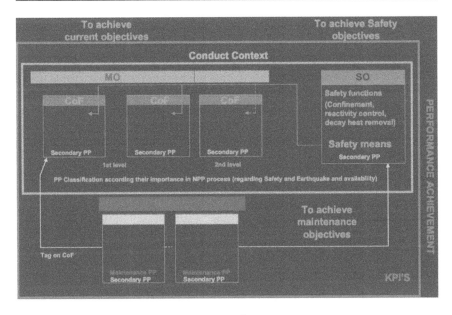

Figure 7.5 Conducting by Objectives concept illustration.

that the COF is necessary but not directly or with a time to act in case of dysfunction.

If the context or the MO is changing, the COF can change its priority level (i.e., means useful to achieve COF may be different with respect to context). The Conducting context gathers MOs and safety objectives while maintenance context defines maintenance objectives with respect to conducting context. The whole part enables building the dashboard using external data to manage performance indicators.

A FA should be performed to support such I&C systems design. Something very important is to allow at any time I&C system to control the context to verify if human operating actions are matching with it. Human operators' actions are observed within the real context (i.e., resulting from an automated PPs analysis – calculated PP) and their situation awareness coming from switches configuration. The final goal is to always display the right information within the right context and with a minimum number of views to avoid excessive browsing among multiple existing views.

Design options have to be chosen. In fact, context is framed by I&C with two switches and four positions. Each position is well defined in the I&C (values, levels of values of PPs, etc.). Regarding the context fixed in I&C by themselves, operators have a frame to operate. Actions are either authorized or forbidden. I&C can block unappropriated actions.

Access to COFs is linked to the context and allows configuring properly displays to operate the process and manage alarms. Alarms are limited to what is just necessary. Displays show only efficient information. Mechanisms of alarms hierarchy follow the same concept and avoid cascading of colors alarms (i.e., *Christmas tree*).

7.8 INFORMATION TECHNOLOGY IMPLEMENTATION

Actually, most of new technologies are not mature enough to be used in complex systems operations with the right level of safety or availability and with respect to mandatory regulations. On the other hand, as Holling, C. S. (1973) proposed, the big issue is resilience and stability of ecological systems. Nevertheless, they could be very helpful on the field or in the operations control room because they are ease of use and provide large amounts of interesting information. As an example, Pad or Laptop can provide operational context-sensitive wireless functionalities capable of simulating any kind of action, providing potential risks preventing human errors anywhere on board, helping in the identification of components using RFID[2] systems, and so on. Another example is the development of connected objects (IoT[3]) able to monitor health of field operators, level of stress, atmosphere conditions in real time, etc. All those parameters are not necessary in a submarine I&C and yet there are important regarding operation and safety and accident prevention or follow up of consequences of an event and contribute to the safety climate.

A submarine is full of narrow compartments. The operation context should be provided in real time to maintain a link between the control room and field operators. Field operators are the eyes and nose of control room operators. This is crucial when both control rooms are at the bow of the submarine. Field operators are part of the shift team or part of the test and maintenance team. Both teams are often in conflict of objectives that the control room must manage at any time with the back-office staff (ashore). The management of co-activities is the crucial issue.

Mobile equipment with smart sensors and video is nowadays essential to monitoring operators, environment, and process at the same time. But we must face new risks. It may be too much easy to install software and mechanisms for cybersecurity are required to avoid malevolent actions. Consequently, dedicated digital I&C display systems should work in parallel. They must be designed with respect to activities with the same concept of operations (CONOPS).

A safety supervision system should provide an overview of submarine operations in any situation at any time, such as:

- status of COFs and Safety Objectives systems;
- trends of the main PPs and margins within safety limits;
- status and availability of safety means;
- alarms hierarchy mechanisms.

A normal-incident supervision system should provide views (e.g., appropriate situation awareness) that fosters appropriate action on the process. These views should be hierarchically organized to support human operator's activity in the right context.

7.9 CONCLUSIONS

More and more, submarines are designed to decrease the number of crew-members to improve the ownership cost with maximum performance and safety. Nowadays, French Ministry of Defense's budgets are decreasing, and designers' approach is cost-oriented. It could remain very expensive during the operational lifecycle. Consequently, the vision of the final product is totally different than in the past, and considerably constrains the implementation of new technologies in submarine design because of the selling cost.

New engineering should try to optimize the safety/availability ratio using an ownership cost approach. Improvement of safety systems is a major key issue. Manual mode is becoming the very last option when it is still possible. This is a big change in the submariner culture, as it was before the only safety mode.

Here are a few HCD recommendations for the future:

- Design should be based on the identification of weak and strong links between functions. Functions should be designed to keep consistent strong links as much as possible when dispatching functions using the concept of Conducting by Objectives.
- At the very beginning, we need to define contexts and think about activities – not about tasks – in the definition of functions. Tasks will come later.
- Design should be driven by trying to cleverly dispatch human-human functions (i.e., considering skills, competencies, responsibilities, and accountabilities of each actor in the overall organization), human-automation functions (i.e., taking into account speed to act, human factors, and so on), and automation-automation functions with respect to the I&C architecture regarding to norms and standards.
- Design should be done using mature technologies (i.e., based on norms and standards for the definition of equipment class levels), taking time to test innovation, and demonstrating the ability to match requirements. Nowadays, simulation enables speeding up these design processes.

As Don Norman said, "Writing as Design, Design as Writing" (Norman, 2018).

New technologies have their place in submarine design and in operation. Without removing norms and standards, the engineering approach should consider available new technologies too to improve human performance. It is not only a question of norms and standards, but it is also a question of human acceptance. Especially here, it is a question of submariners/specific workers acceptance in link with the job rules fully matching with submariners' culture. Young people will operate future submarine with better knowledge of new technologies. Older operators could help in the operational implementation of these new technologies. Considering the iPhone experience, nobody could think a few years ago that Smartphones could flood the market at it does.

The speed of submarine engineering is slower than other current new Information Technologies. This is the reason why it is necessary to build a concept of operation living independently from the two engineering lifecycles. The level of abstraction of the concept needs to have a chance to implement concepts within last smart components. It is mandatory to allow operation experts working with designers.

NOTES

1 In normal and incidental situation, Conducting Function are Functions to pilot.
In accidental situation, SF are functions to pilot. The conducting function remains as an extensive means to reach Safety Objectives if they are still working.
2 Radio-frequency identification devices.
3 Internet of things.

REFERENCES

Boy, G.A. (2008). *Cognitive Function Analysis*. Greenwood/Ablex, Westport, CT.
Boy, G.A. & Brachet, G. (2010). *Risk Taking*. Dossier of the Air and Space Academy, Toulouse, France, ISBN 2-913331-47-5.
Holling, C. S. (1973). Resilience and stability of ecological systems. *Annual Review of Ecology and Systematics*, 4, pp. 1–23.
Loine, L. (2014a). *Risk Taking in Managing a Nuclear Submarine*. International Workshop on Risk Taking, FIT, USA.
Loine, L. (2014b). A solution of an appropriate HMI Design concept for NPP – Links with Functional Analysis and Information Technologies. *AHFE 2014*. Jagiellonian University, Kraków, Poland.
Norman, D.A. (2018). Chapter 17: Writing as Design, Design as Writing. Retrieved from the Web on March 4, 2022, https://jnd.org/chapter_17_writing_as_design_design_as_writing/

Chapter 8

Risk mitigation practices in commercial aviation

De Vere Michael Kiss
Consultant, expert, and teacher pilot
Melbourne, Florida

CONTENTS

DOI: 10.1201/9781003221609-8

8.1   INTRODUCTION

Since I was a young boy, I dreamed of becoming an airline pilot. Little did I realize that dream would become a pathway to higher education, interactions with some of the most intelligent individuals, and a lifelong journey of self-improvement. My aviation career began in 1983 when Pacific Southwest Airlines (PSA) hired me as a flight attendant. That job provided the economic pathway of becoming an Airline Transport Pilot (ATP) and a Doctor of Philosophy (Ph.D.). Since 1983, I have worked for six airlines; through mergers and acquisitions; as a flight attendant, first officer, captain, and instructor. Additionally, my education has endowed several degrees: two associate degrees, a Bachelor in Aeronautics, a Masters in Human Factors, a Masters in Aviation Space Sciences, and a Ph.D. in Human Centered Design. Additionally, I have taken and taught several advanced courses in psychology, biology, microbiology, human anatomy and physiology, human factors, crew resource management (CRM), aviation law, safety, and accident investigation.

For me, personal motivation to participate in risk mitigation practices began after reading the National Transportation Safety Board (NTSB) reports of several commercial airline accidents that occurred during the 1970s. Those reports discovered that many of the accidents occurred as a result of a dysfunctional safety culture that existed at the time; a machismo culture where superiors had supreme authority and subordinates had difficulty expressing pertinent information. The findings and recommendations of those NTSB reports led to the enhanced robust culture of today; the team dynamic known as Crew Resource Management or CRM.

8.2   MITIGATING RISK IN AIRLINE OPERATIONS

When applied to aviation, there are a number of risk variables involved when managing a complex modern commercial aircraft: weather, terrain, traffic, systems, systems interaction, automation, crew coordination, human-machine-interaction (HMI), security, maintenance, communication, and many others. Airline operations and procedures are dynamic, not linear; therefore, it is difficult to use probability statistics to mitigate the complicated risks involved.

The current airline accident rate is low, one in every one million departures (Boeing, 2012). This fact advocates that the industry, in its entirety, is working very hard to mitigate risk. However, because the magnitude of the consequences of just one mistake can be catastrophic, risk is an element that cannot be ignored. Therefore, continual assessment is required to reduce the probability of an accident occurring, particularly as more flights are added to the system, i.e., with more exposure, the probability of an accident or incident becomes increasingly likely.

Wood tells us that increasing the number of flights will add an additional exposure of risk to each flight (Wood, 2003). According to the Federal Aviation Administration (FAA), the average number of aircraft in U.S. airspace, at peak hours, is 5,000. Further, there are, on average, 44,000 daily flights within the United States (FAA, 2020); this is a total of more than 16,100,000 flights a year. Considering these numbers statistically, the potential of risk is no longer low. Thus, exposure is the biggest single factor when measuring the probability and/or magnitude of risk (Wood, 2003). Hence, the element of risk must be measured and a determination made as to whether the risk involved is acceptable.

When assessing risk, one must recognize a "potential risk hazard" (a present condition, event, object, or circumstance that contribute to an unplanned event) when starting the risk management process. The FAA's Risk Management Handbook (RMH) elucidates that the ability for one to recognize, predict, and project a hazard into the future is based on that individual's personality, education, and experience. I agree with this assessment and will point out several methods used in the airline industry that help pilots manage risk in aircraft operations.

Because it is difficult to quantify data concerning how individuals assess and react to risk; as each person evaluates situations differently; we cannot adequately predict how individuals will perform in certain situations, and, therefore, must elicit, with strong convictions, standard operating procedures (SOPs), and policies, such as CRM or multi-crew coordination (MCC), that include hazard and risk evaluation and, moreover, the required actions to mitigate and lower risk.

8.2.1 Risk evaluation

When designing and operating sociotechnical systems, it is necessary to describe and develop scenarios that can be validated and tested during modeling and simulation. This is desirable when defining the scope of situated risks and potential solutions, i.e., scenario-based design. This is to establish risk models.

To create a prevention path, one must identify a potential risk. Additionally, if a prevention path is established and the event still occurs, a recovery phase to reduce the consequences of the emergent property is required. This is accomplished by incorporating resilience engineering during the design stage and continuing it through production and the life-cycle of the artifact (Boy, 2011; Hollnagel, Woods, and Leveson, 2010; Hollnagel, Pariés, Woods, and Wreathall, 2011). Additionally, procedural engineering and training should be involved to generate measures for reducing the severity of risks, further reducing the consequences.

8.2.2 Risk mitigation evolution

During WWII through the 1970s, pilot training philosophies were mainly concerned with individual performance of pilots. Training the error out of pilots was the major component. However, psychologists realized that because technologies were advancing so fast, it was difficult for the human operator to keep up with the technological changes.

Throughout the 1970s, there were 138 worldwide airline accidents (Appx 1.15 per month). Socially, the air safety record was unacceptable. As a result, it became obvious there was a flaw in the philosophy of training error out of the individual. Therefore, human factors, CRM, and team skills were targeted as the primary considerations for enhancing technical and procedural proficiency.

Seventy percent of airline accidents are attributed to human error (HE). Seventy percent of those HE accidents are directly related to some form of communication problem, difficulties between flight crews, between flight crews and Air Traffic Control (ATC), the interpretation of material in the flight operations manual (FOM), communications between the manufacturer and the user, etc. Additionally, the culture of the time was one that gave the captain supreme authority and he/she was never to be questioned. This was found to be a causal element in many accidents.

In December 1972, Eastern Airlines flight 401 crashed into the Florida Everglades because the captain was distracted by a faulty landing gear warning light and did not maintain proper oversight of the aircraft's flight trajectory as he failed to ensure that one of the flight crew members monitor flight instruments (NTSB, 1972). In 1977, a KLM 747 and a Pan Am 747 collided on the runway at Tenerife Airport in the Canary Islands because the KLM captain's judgment was influenced by the need to get back to his base of operations. In December of 1978, United Airlines flight 173, a DC-8, experienced the loss of all four engines, due to fuel starvation, and crashed several miles short of the runway in Portland Oregon. The NTSB cited the captain's inability to listen to other flight crew members and the failure of those other flight crew members to communicate more directly to the captain as causal elements for the accident (NTSB, 1979). As a result of these accidents, the NTSB made recommendations that airlines provide participative management classes for captains and assertiveness training for other cockpit crew members (Kanki, Helmerich, and Anca, 2010). This was the beginning of CRM; Also known as MCC in Europe.

8.3 CREW RESOURCE MANAGEMENT (CRM)

Prior to the 1980s, flight crews did not perform as a team, and subordinate crewmembers did not generally assert any authority while in flight. This normally resulted in crewmembers waiting for an order from the captain before performing any task, and inhibited crewmembers when responding to abnormal or emergency situations.

With the discovery of this dysfunctional culture, it became apparent there was a need to change the training philosophy from one of individual performance, to one that incorporated all team members associated with the flight. There were five evolutionary generational iterations involved in the creation and development of the CRM/MCC approach utilized today (Helmreich, Merritt, and Wilhelm, 1999).

CRM was first adopted in 1981 by United Airlines following a 1979 NASA workshop promoting resource management on the flight deck. The workshop focused on changing the training philosophy to a group/team system, i.e., the captain should include the other flight crew members.

Another workshop goal was to change the management's organizational structure to include a corporate culture that created and supported SOPs in dealing with normal, abnormal, and emergency scenarios. Furthermore, another goal was to enhance the practice of quick identification and resolution practices through procedural training to augment situation awareness (SA) and decision-making (DM). Finally, the goal was to incorporate methods to evaluate training and line operations to ensure CRM was working and make changes when warranted.

A second CRM workshop took place in 1986 and focused on embedded CRM methods in the fabric of flight training and flight operations (modeling and simulation). This included team building concepts via enhanced crew briefing strategies, stress management, DM strategies, and ways to break potential chains of errors.

The third generation of CRM emerged in the early 1990s and was broadened in scope via reflected characteristics of the aviation system and how crews functioned within it, and the integration of CRM with technical training. This included training to teach pilots specific skills and behaviors to help crews function more effectively using CRM and flight deck automation issues.

The fourth iteration of CRM was also produced in the early 1990s and addressed the Advanced Qualification Program (AQP); a voluntary and flexible training program designed for specific needs of the organization including: utilizing CRM during line oriented flight training (LOFT); the formal evaluation of crews in full mission simulation; i.e., human-in-the-loop simulation (HITLS).

The fifth generation of CRM attempted to get everyone on board and developed new CRM countermeasures including error avoidance, the trapping of incipient errors, and mitigating error consequences. It should be noted that the fifth generation of CRM also included the in-flight crew (flight attendants) in training.

Today, CRM accepts that HE is ubiquitous and, therefore, inevitable. Accordingly, CRM serves as countermeasures for HE. CRM incorporates three lines of defense (Kanki, Helmerich, and Anca, 2010).

1. Error avoidance (SOPs)
2. Trapping incipient errors before they occur (an input-output process)
3. Mitigation of the consequences of errors once they were committed

As an airline pilot, my interpretation of CRM is the focus on proper responses to potential threats of safety and, ultimately, the proper management of crew error. CRM requires that the flight crew-/team members reach beyond the evaluation of individuals and concentrate on the entire team responsible for the safety of a flight. To achieve this, there was a need to develop enhanced communication methods between every individual involved with an individual flight. This also required redefining some terms (in regard to aviation): key team members and team members.

In the context of CRM, everyone who participates in a flight is a team member. Key team members are those who are physically on board the aircraft: the flight crew, automation, in-flight crew, and passengers. CRM requires effective communication between key team members and the team members outside the onboard aircraft environment. One of the most effective types of verbal communication consists of crew briefings.

The crew briefing is accomplished at the beginning of the trip (the airline term for a schedule of flights a crew will perform from one to four days). It is the captain's responsibility to create a positive team building milieu. This is most important when a crew has never flown together before. This is accomplished by conversing with the key team members in a private setting before boarding passengers (the passengers receive their briefing during the flight attendant emergency demo).

Crew briefings are necessary to clarify the individual task responsibilities of each crew member and the environment in which the flight will be conducted (Helmreich, Merritt, and Wilhelm, 1999), i.e., an open communication and team-oriented setting. The crew briefing should ensure the entire crew understands the roles and responsibilities expected of them.

The current focus of CRM is on the attitude, motivation, behavior, and performance of individuals in a group setting (Curtis and Jentsch, 2010). A positive briefing by a captain consists of an atmosphere were crew members feel comfortable to speak up, state opinions, ask questions, and challenge when necessary (Helmreich, Merritt, and Wilhelm, 1999). Individually, I also explain to the crew that proficiency is the hallmark of an effective team as described by Tullo (2010) and that it is important for each team member to be proficient at the task(s) assigned to them.

During the briefing, the leader must evaluate that each member is physically and mentally healthy; and competent in preforming their shared authority (their responsibility for certain functions). If the captain discovers that a team member is lacking in health or the understanding of their roles, he/she should have that crew member replaced. The team's overall performance can be degraded when a member is impaired or lacking the necessary knowledge to perform their duties.

I learned this on my very first day as an official first officer after completing my simulator training and initial operating experience (IOE) training. During the engine-start-phase of the first of four engines at a Las Vegas gate, an engine fire occurred and an emergency was declared, requiring an emergency evacuation of the passengers from the aircraft, a BAe-146. The captain and I began our memory items when I looked through the flight station doorway and noticed that passengers were departing the aircraft through the right forward service door.

When I gave the emergency evacuation command over the public address (PA) system, I explicitly stated, "Left side only". This was due to the fact that the fire was on the right side of the aircraft. When I informed the captain that passengers were deplaning on the right side of the aircraft, he made the decision and directed me to leave the flight station and perform the flight attendant's duties. This was because I had previously served as a flight attendant for the company and he knew that I was capable of accomplishing the flight attendant's duties. However, after I left the flight station, the captain had to complete his duties as well as mine.

In this situation, the flight attendant did not perform her duties as directed. In fact, she was frozen against the bulkhead as people ran by her. I had to physically put my left arm across the exit door and use my right arm to point to the left door and verbally yell, "Go this way, go this way"! Therefore, the captain and I had to redelegate our responsibilities to achieve another crew member's shared authority. All passengers were safely evacuated and no one was injured. The captain successfully completed all required checklists and after the event was over, we, the captain and I, accomplished a successful three engine ferry-flight to San Diego for maintenance.

Note: This was an isolated incident and not intended to stereotype any individual. In my four-decade career, I never experienced another issue with a team member not performing their duties correctly.

The point is that when one team member cannot function properly, the probability of a successful outcome during an emergency is lowered and, therefore, risk is increased. This is unacceptable in a life critical environment and why all team members must be able to perform their assigned duties.

During the briefing, it is important for the captain to positively motivate those under his charge. I accomplish this by assuring the team of my competence, and the skills I developed through the company's training. However, I also tell the team that I am not perfect and, in fact, vulnerable to error; that I need them to help prevent any potential error and that any information they can provide will increase my SA, and their input is vitally important to me.

Expressing that they are important team members and that I value their input, generally increases communication between the cabin crew and the flight crew. Establishing good CRM requires positive leadership (Helmreich, Merritt, and Wilhelm, 1999; Curtis and Jentsch, 2010). Whether a crew works well together or not is a function of the captain (Tullo, 2010). The captain's tone and manner influence the openness or alienation of the team.

In my experience as a simulator instructor, and acting as a captain, when a captain recognizes a problem as a crew problem, and manages the problem accordingly, the crew typically does well. However, if the captain handles the problem individually, the crews do not fare so well. Therefore, the captain's behavior carries considerable weight in how well a crew performs their responsibilities, affecting the overall teamwork and organizational aspects (Tullo, 2010).

Additionally, the leader should discuss the range of shared authority in which groups most effectively operate, i.e., everyone knows who will do what and when they will do it! Furthermore, captains need to demonstrate their authority in three ways (Tullo, 2010):

1. Briefings should be organized in a logical parameter
2. Briefings should always contain nomenclature specific to aviation
3. Ensure the crew that he/she is comfortable leading the group

I would also add that a good leader must be able to switch from democratic and autocratic authority. During normal operations, a leader can be democratic and enlist support in a friendly and charismatic format. During abnormal operations, if time permits, a leader can ask for suggestions and decide in a democratic presentation with a mixture of autocratic demands. However, during an emergency, or when time compression occurs, a leader's decisional speed requires an autocratic style and the team should behave accordingly, unless they are aware of pertinent information that the leader does not have.

The global airline accident rate has decreased since CRM was introduced. CRM has fundamentally changed the social cultures of the industry including regulatory entities, management, unions, and the individual professional cultures existing within those groups. More importantly, CRM has helped to improve the way the industry deals with threats and errors that diminish safety.

8.4 THREAT AND ERROR MITIGATION (TEM)

Typically, the pilot is the individual often blamed for an aviation accident (Hawkins, 1997; Salas and Maurino, 2010). However, if the investigators look deep enough, organizational and procedural immaturity has been found to be responsible human factors linked to the actions of the pilot, i.e., the pilot was performing within the policies and procedures outlined by the company, manufacturer, and/or regulatory entity (Leveson, 2011).

When accidents happen, it is necessary to explore all of the associated agencies to determine what went wrong so that procedures and training can be developed to reduce the risk when unsafe events express themselves (Hollnagel, Woods, and Leveson, 2010; Hollnagel, Pariés, and Woods, 2011; Leveson, 2011). According to Hawkins, there are three basic tenets with respect to human error (Hawkins, 1997):

1. The origins of errors can be fundamentally different
2. Anyone (even airline captains) can make errors
3. The consequences of similar errors can be quite different

Recognition of these three tenets is an essential basis for making progress toward the mitigation of risks within airline operations. Any individual placed into a simulator, given enough negative variables, can be engaged into a precarious situation that would be difficult for even the best of aviators to recover.

According to Baldwin, when people function near the edge of their operating envelope, they can be cognitively limited because they have exhausted their reserve capacity of resources (Baldwin, 2012). Thus, increasing task demand further can result in the operator being overloaded, resulting in a reduction in performance, i.e., decisional speed and accuracy can be impaired.

One of the methods developed for reducing HE is threat and error mitigation (TEM). TEM was incorporated with modeling and simulation, scenario-based training, and CRM. TEM is used to instruct pilots, during simulation training, to work as a team and reward team members when they commit an error, identify it, correct it, and verify it.

An error is something that is committed (intentionally or not) by any crew member in the flight station. It is imperative that each crew member be proficient in monitoring the other, and also monitor the automation and aircraft systems; and verify that all are compliant with the desired state of positive stability. There are three entities in today's flight station: 1. the captain, 2. the first officer, and 3. the automation (flight management system [FMS], autopilot [AP], warnings, GPS, etc.).

To recognize errors in the flight station, I utilize what I call, "Critical Triangles of Agreement" or CTA. This means the key team members within the flight station must agree. If one does not agree, the crew must recognize there is a possible error and take appropriate action. For me, appropriate action is to recognize that one of the three entities is *not* in agreement, and the current path needs to be altered by asking ATC for radar vectors or a holding pattern; this affords the necessary time to resolve the conflict and ensure that the flight crew is in agreement. When this occurs in the simulator, the crew is rewarded and, accordingly, the crew should function in such a manner in the "real-world". For the reader, this requires some definitions so the reader can understand the dynamics of a flight crew. The captain is the sole person responsible for the safety of a flight. He is always in command. The first officer is second in command. The captain and first officer reverse flying roles that require further definition.

In flight, the pilot flying, or PF, is responsible for manipulating the flight controls, engaging the FMS, and AP when using the automation and calling for appropriate checklists. The pilot monitoring, or PM, is responsible for manipulating the FMS and AP when the PF is manually flying the aircraft (i.e., flying the aircraft by hand), reading the checklists and verifying checklist compliance, talking on the radio to ATC, and monitoring the PF. There are rules that enhance the synergies of the PF and the PM.

The PM must be efficient at monitoring and performing duties of the SOP's and duties assigned by the PF (Tullo, 2010). The primary job of the PM is to monitor the progress of the flight, the PF's performance, the automation, and inform the PF when he/she detects any threat or error that can lead to negative consequences. Conversely, the PF must also monitor the automation and inform the PM when he/she detects a threat, error, or deviation from normal operations. If a threat or error is detected that crew member's job is to make an assertive challenge that will identify the error so it does not occur, or is corrected.

This requires several skills (Tullo, 2010):

- Vigilance: a very important skill when monitoring and, even more important, the individual should know when to be vigilant (keep in mind, no one can attain this all the time and there must be periods of relaxation in between periods of vigilance)
- Envisioning: must stay ahead of the aircraft.

Note: I call this envisioning, "Mental Simulation" or "Mental Modeling". The pilot must mentally project the aircraft at some distant point ahead of the current aircraft trajectory and verify the aircraft will arrive on course, on altitude, on geographic location, and on time. This skill develops a plan for the entire crew and is an absolute necessity for good SA, DM, and preventing complacency; especially when using automation. This act lowers risk and keeps the pilots at a level of awareness that allows them to react at the earliest moment; the sooner a pilot recognizes a problem and corrects it, the higher the probability of a positive outcome. Conversely, the longer it takes for problem recognition to occur, the higher the probability of a negative outcome. We call this "Recognition and Response".

- Communication: the PF must communicate their thoughts, evaluations, and envisioning to the PM and verify his/her plans are in agreement with the PM and the automation. This establishes CTA and enhances accurate hypotheses and reduces continued paths of error (crew fails to recognize they have a false hypothesis of the true situation)
- Adaptability: the ability to adjust to changes; an absolute necessity in crew dynamics
  - "Decision bias" and "plan continuation error" have been cited as causal factors in many aircraft accidents over the years
- Receptiveness: the ability to pay attention to the other pilot's idea's, concerns, and or questions

Receptiveness, adaptability, and a willingness to change are key elements of a safe flight station (Tullo, 2010). When one uses logic and tact, a leader/follower can inspire the other crew member to commit to ideas or actions. The key is to make the information transparent to the other crew member.

Additionally, crews are trained to recognize threats. What is a threat? A threat is anything that occurs outside of the aircraft that may jeopardize the safety of the flight. This can be another aircraft, or a mistake made by a controller and the crew must recognize the threat and respond accordingly.

8.5 STANDARD OPERATING PROCEDURES, PERFORMANCE AND REQUIRED KNOWLEDGE

In airline operations, mitigation of risk is accomplished using SOP's, CRM, TEM, simulator training, the LOFT, check rides, line-checks and other approaches. In addition, training should include system and performance information and procedures within proper documentation.

8.5.1 Information/documentation

It is important that operating manuals transfer detailed information to the flight crews. This is because ambiguous information can lead to subjective inference on the part of the operator. The vaguer something is, the more subjective it is. Subjective inference leads to uncertainty, and, therefore, the probability of a positive outcome is lowered. Accordingly, the knowledge transference of objective instruction for conflict resolution must be clear and concise.

The information must include normal, abnormal, and emergency SOPs, checklists, and amplified instructions that designate who will accomplish the tasks and when those tasks will be accomplished. Additionally, responsibility is mandated through what is termed "shared authority". Shared authority describes who has responsibility for certain functions during the different phases of flight, i.e., the captain, first officer, and automation share duties and the captain can delegate those responsibilities as needed.

8.5.2 Regulations and SOPs

In the United States, the FAA has developed the Federal Aviation Regulations (FARs) to list who is responsible for specific actions and how those actions are performed. However, there are times when an unexpected/unanticipated event may occur, and no clear and concise documentation exists for the handling of such an event. This is when creative thinking must override practice and procedure to regain stability. In such an event, the FAA gives the crew the authority to deviate from any FAR, or SOP, to meet the extent of the emergency or unexpected/unanticipated event.

Once initial SOPs are established, they are practiced during HITLS and perfected through an iterative process and then presented to the users through the appropriate manuals. The users then practice the procedures (procedural training) during simulator training and must qualify for revenue operations through an evaluation by an FAA designated check airman.

8.5.3 Procedural training

Quality training must exist for the operators to address potential risk situations correctly and in a timely manner. If the training is of low quality, risk mitigation will be low. Without proper training, operators are susceptible to a lack of knowledge that can cause crewmembers to become situationally unaware, such practice leads to uncertainty.

Most airlines provide very good training; however, some carriers do attempt to lower costs through a reduction in training. My feeling is that all airlines should be regulated and required to incorporate the same level of training, at any cost.

Within the airline domain, there are two types of training:

1. Systems training (declarative knowledge) includes:
   - Systems
   - Systems integration

2. Procedural training (procedural knowledge) includes:
   - Normal flight ops
   - Abnormal flight ops
   - Emergency flight ops

Systems training occurs in the classroom and during home study.

Procedural training occurs in several different training artifacts:

- Home mockup: cockpit panel posters
- Flight-training device (non-motion)
- Full-motion 3-axis flight simulator
- Computer programs

During simulator training, pilots continually practice scenarios that help them develop and enhance their expert skills. This enhances their decisional speed and accuracy. Further, modeling and simulation training can influence and improve advantageous behavior during emergencies, affording pilots with improved abilities to regain positive stability (Hollnagel, Woods, and Leveson, 2010; Hollnagel, Pariés, and Woods, 2011; Leveson, 2011). This occurs because practice helps pilots to act in a reflexive, natural, and instinctual manner (Ericsson, Krampe, and Tesch-Romer, 1993; Nokes, Schunn, and Chin, 2010).

For a pilot candidate to transition to the simulator and begin procedural training, they must complete their systems training. Systems training prepares the candidate to begin acquiring the declarative knowledge required to succeed in the simulator. Without systems and systems interaction knowledge, it is impossible for the pilot in training to succeed in the simulator because they lack the deep systems knowledge essential for performing required procedures adequately.

8.5.4 Training scenarios

Training should also include events that involve past accidents so that individuals have knowledge of the events that led up to those accidents, providing them with experience that will help to prevent future occurrences of the same accident. I find that my best risk mitigation practices come from studying aviation accidents without judgment. The key is to put yourself in the situation and ask, "what can I do to prevent this from happening to me, or, if it does happen, how can I reduce the magnitude of the consequences of such a scenario"?

To facilitate the above, a clear understanding of the types of risk should also be included during pilot training. By incorporating types of risk with training, a systematic process of DM can be transferred to the willing pilot, which should lead to proper risk assessment. This DM process is designed to systematically identify hazards, assess the degree of risks, and determine the course of action (RMH, 2016). Thus, proper training scenarios are incorporated to facilitate these operator necessities.

8.5.5 Preflight planning

Preflight planning is when hazards/risk and potential mitigation should be evaluated, identified, and either reduced, or, if possible, eliminated. That is the time to consider weather at the departure, enroute, and arrival.

Additionally, traffic congestion, terrain, aircraft performance, and other potential threats should be considered and evaluated. Preflight planning allows a pilot to manage some hazards in a preemptive manner to avoid risks. One practice I perform for preflight planning begins the night before; I review every chart, approach plate, and company special procedures for every airport I will be operating in and out of. This requires highlighting and noting every element that will be incorporated into my crew, departure, and arrival briefings.

8.5.6 Manuals and checklist usage, the products of SOPs

To enhance CRM, a company must provide clear and concise documentation required for every conceivable scenario performed during normal, abnormal, and emergency conditions. Using standardized procedures and terminology promotes crew coordination, communication, understanding, and *verification* practices. This also requires learning industry nomenclature. Proper use of the Normal Procedures Checklist reduces unsafe practices, complacency, carelessness, and the development of individualized procedures as it standardizes user actions, i.e., who will do what and when they will do it.

8.5.7 Normal procedures

The Normal Procedures Checklist is designed to be quickly and easily accomplished in logical time sequence during a flight. Checklist groupings are selected so the items are consistent with established flow patterns (the physical location of individual switches, buttons, and levers in the flight station).

The Normal Procedures Checklists are comprehensively developed for individual flight phases. These checklists are not a do list; rather a verification list (each pilot must look and verify the item is accomplished) to ensure essential checks and procedures for that particular flight phase are accomplished.

The challenge and response concept:

- Airline industry checklists incorporate the challenge and response concept:
  - Challenger: the person who reads the checklist and verbally issues the challenge
  - Responder: the person who verbally issues the response

- Two basic tenets:
  - The challenger will call aloud and visually verify that each action matches the correct checklist response
  - The responder also visually verifies that each action matches the correct checklist response and he/she will also give a verbal response

8.5.8 Flows

In addition to checklists, the industry has established what are termed "Flows". Flows are another important element in the reduction of risk. As stated above, the use of checklists is based on the "Challenge and Response" philosophy. This is to incorporate a dynamic of checks and balances for verification. However, there is a third level of redundancy to improve safety further.

Each checklist has an associated flow. A flow is a memory item that parallels the physical movements of the crewmembers' motions in accomplishing a particular checklist. The responsible crew member accomplishes the flow by memory. When the captain, or the PF, calls for the checklist, the challenger (PM) reads the checklist out loud and the responder (PF) responds, followed by the PM also responding out loud. Again, both pilots must look at the specific item and visually verify correct position or status before responding. Thus, each item on a checklist is checked a minimum of three times, reducing risk and enhancing safety.

8.5.9 Checklist amplification

Normal cockpit checks and operating procedures are presented in the expanded checklist located in the aircraft operating manual (AOM) or the flight crew operating manual (FCOM). Where applicable and appropriate, checklist amplification is used to spell out particular procedures/policies.

8.5.10 Abnormal and emergency procedures

Abnormal and Emergency Procedures Checklists are incorporated only after repeated testing and investigation, i.e., iterations of HITLS. They represent the best-known available facts about the particular subject. Pilots must use these procedures as long as they fit the abnormal condition. However, if at any time they are not adequate or do not apply, the captain's best judgment shall prevail.

All crewmembers must be thoroughly familiar with abnormal procedures and equipped to handle the duties of any other crew member. Abnormal checklists are used in the same challenge and response method as the normal checklists; with one additional step: the pilot reading the checklist (normally the PM) should read aloud both the challenge and the response. There must be no doubt in any flight crewmember's mind as to the correct course of action. The pilot responding has the same responsibility for checking and/or accomplishing the item.

8.5.11 Un-planned-for events

Checklists have been designed for scenarios that are planned for. However, there are times when an un-planned-for event will express itself. During such an event, pilots must be creative in their thinking in order to mitigate the magnitude of risk associated with an un-planned-for scenario. For an individual to operate a modern aircraft and manage the associated systems during an unusual event, they must have a deep knowledge and understanding of how the systems work separately and how the failure of one system will affect all other systems and, ultimately, the performance of the aircraft.

8.6 VARIABLES INVOLVED WITHIN AIRLINE OPERATIONS

There are numerous variables within the complex environment of airline operations. Such variables must be considered and understood. In this section, I describe many of the variables I have discovered during my career as an airline pilot, instructor, educator, and designer.

8.6.1 Deep knowledge

Today's aircraft are very complex machines and require the pilot to have a deep understanding of their operations. It is particularly important for the operator to understand how aircraft systems interact with one another, especially during system failures. Telfer tells us that when an individual is acquiring such knowledge, their success in a relatively difficult task requires that they relate new knowledge with old knowledge (Telfer, 1993). According to Telfer, pilots acquire deep knowledge in several ways (Telfer, 1993):

- Search more information about a topic
- Absorbed in learning the topic
- Want to understand the information completely
- Focus on:
  - Problem-solving
  - Judgment
  - Decision-making

It has been my experience that the attributes Telfer describes are, indeed, valid. However, I would add that the individual must also have a heightened sense of SA to enhance such characteristics. In my view, without proper SA, a pilot candidate will be limited in their problem-solving, judgment, and DM.

Thus, airline pilots must have a deep understanding to have enhanced SA. Pilots are extremely proficient at adjusting and coping with changing environments and changing circumstances. They are "surface-oriented-learners" who are well trained at the following (Telfer, 1993):

- Setting personal goals
- Reviewing material (retrieving necessary information quickly)

- Testing their individual current knowledge about:
  - Systems
  - Procedures
  - Aircraft performance
- Developing summaries to update their current knowledge

I would add one more learning trait to those considered to be surface-oriented-learners:

- Formulating more than one resolution for a given scenario and recognizing which one will best address the possible situation

Those of us in the airline industry describe this activity in another way; airline pilots develop solutions to complex problems by asking themselves "what-if" questions and then search system information to find solutions to those what-if questions. This is how pilots develop their deep knowledge, expertise and skills of enhanced DM, and problem-solving methods.

In my experience, I have often marveled at how pilots have found solutions to unexpected scenarios given in the simulator during training. Pilots often confer with one another about potential solutions to a specific what-if question they could not personally resolve. In this way, there is a consensus among the team and a solution is collectively found (i.e., these individuals develop deep knowledge by asking specific what-if questions and then find a solution, individually or collectively).

8.6.2 Developing positive knowledge schemas

Through the repetition of simulator training, assimilation and accommodation allow pilots to instill schema patterns that permit them to perceive, process, decide, and act in a reflexive behavior pattern (Rathus, 2008). Jean Piaget (cognitive psychologist, 1896–1980) believed that "true intelligence" involves adapting to the world through a smooth and fluid balancing of assimilation and accommodation processes (Rathus, 2008).

Piaget's accommodation mechanism states that knowledge accumulates in memory following an iterative process. Therefore, through the actions developed during procedural training, positive knowledge schemas are developed which can be articulated and reshaped to enhance emergent anomaly resilience and robustness. This is particularly important in respect to temporal limitations during an emergency where high workload and limited available time (time compression) are a factor (Connor, 1985).

When time compression occurs, pilots must act instinctually, intuitively, proficiently, and accurately; to mitigate the emergence of an unwanted event, enhancing resilience and robustness against an unstable situation (Hollnagel, Woods, and Leveson, 2010; Hollnagel, Pariés, and Woods, 2011; Leveson, 2011).

8.6.3 High workload/short temporal component (time compression)

Different people have different levels of experience, education, and skills; and different ways of perceiving, processing, deciding, and acting on information. Thus, there are many complex sources of variation in workload demand (Telfer and Biggs, 1988; Jensen, 1995; Baldwin, 2012). Workload demand fluctuates with the diverse task functions that occur over time and during the different phases of flight, i.e., takeoff and landing versus cruise flight.

Workload is further complicated by the operator's personality and the way they respond to tasks or personal demands (Salas, Bowers, and Edens, 2001). This response can be further affected singularly or in combination with multiple tasks, which increase the challenge of the agent to properly assess workload.

Multidimensional influences can inspire disassociation between the tasks and required actions. This can lead to an overload situation that affects the person's cognitive abilities (Connor, 1985; Jensen, 1995; Hawkins, 1997; Dismukes, Benjamin, and Loukopoulous, 2007). Therefore, normal DM performance can deteriorate when workload is too high or available time is critical. During such times, it is important for the individual to let the other pilot know that a high stress situation exists, and to delegate tasks appropriately. Operational policy must allow for the delegation of activities accordingly when an overload situation exists. This is when CTAs are critical and the crew needs to identify stress or disagreement between the three flight station entities and take appropriate action, i.e., request radar vectors or a holding pattern until all three entities agree, or, if agreement cannot be reached, ask for help from ATC or any other associated team member.

8.6.4 Decisional speed and accuracy

As a pilot enters the approach phase of flight; time compression and workload are increased. Fifty-four percent of accidents occur during the approach to landing phase of flight (Boeing, 2012). As workload elevates, time compression exponentially becomes a factor; consequently, workload parameters require enhanced decisional speed and accuracy. This requires continuous rehearsal of procedural training that affords pilots with procedural memory (knowledge schemas) to augment their decisional speed and accuracy. This enhances the pilot's ability to instantly recall critical information and act quickly and accurately, enhancing robustness and resilience to unwanted events (i.e., this is intuitive resolution for DM processes) (Connor, 1985).

8.6.5 High stress (emotion)

When a pilot experiences an anomaly during flight, the sympathetic nervous system (SNS) automatically induces hormonal activity that will (Chappelow, 2006):

- Increase heart rate
- Increase blood pressure
- Reduce blood flow to internal organs
- Reduce cognition
- Activate other physiological changes

I have personally experienced this physiological phenomenon during in-flight emergencies and I can state, unequivocally, that it takes considerable effort to ignore the associated symptoms in the effort to manage the situation at hand. Gawron declares that emotions can reduce a pilot's ability to resist spatial disorientation (Gawron, 2004).

Accordingly, SNS activation can degrade user performance. Through practice and adaptation, the user will utilize his/her automatic mental representations to adjust to SNS cognitive dysfunction. However, these factors can lead to reliance on automatic processing, repetition, and habit capture, and expectation.

8.6.6 Automatic processing, repetition, habit capture, and expectation

"Automatic processing" (highly practiced skills that become automatic over time) is mentally established during the repetition of training over a long period of time. Automatic processing has the advantage of being fast, efficient, and can require minimal demands on cognitive resources (Salas and Maurino, 2010). However, Salas and Maurino discovered that while automatic processing is ordinarily very reliable, it can become less reliable when there is a change in normalcy (i.e., a person's attention can become diverted when initiating a task) (Salas and Maurino, 2010).

When individuals intend to divert from a normal step in a habitual procedure, they are vulnerable to what is known as "habit capture" (Salas and Maurino, 2010). Habit capture occurs when a person has repeated the same task over and over many times, and they acclimatize to the same procedure. However, when there is a change in the procedure, they unconsciously perform the old procedure. If a pilot does not pay careful attention, they may be in danger of reverting to the habitual step rather than performing the intended and correct one.

"Expectation" occurs when an individual is expecting a certain outcome based on previous experience. An example would be, setting an old decision height altitude during an instrument landing when the altitude was changed during a revision. Such a scenario could place the aircraft in danger of hitting an object on the ground prior to landing. Another example would be expecting the same taxi clearance assigned during the last flight, but the controller gave different instructions.

Therefore, automatic processing, repetition, habit capture, and expectation can influence an individual to revert to actions they have procedurally practiced during an emergency, when the action is no longer warranted. Consequently, one must understand these potential situations and be mindful not to be affected.

8.6.7 Cognitive unitization, or chunking

The process of developing expert cognition tools induces unitization that occurs when new neural encoding and neural processing cause normal cognitive components, once perceived separately, to fuse together (Dror, 2011). This creates a new schema that can be used later and is known as "chunking" or "lumping". This is how neural processing provides cognitive optimization for expert cognitive information processing; this is needed for expert domain performance (Connor, 1985). For example, chunking occurs when one fuses a list of items into one memory item. An example would be taking seven items and remembering them as one item. An average individual can recall about seven items or digits.

The average airline pilot can recall anywhere from 9 to 11. This is useful because an individual can incorporate seven items that include a chunked set of seven items and, theoretically, recall up to 49 items or digits instead of just seven. Pilots have developed this skill is very useful for analyzing and troubleshooting unforeseen abnormal situations. It is particularly advantageous during high workload and time compression when decisional speed and accuracy are required for safe resolution to an unwanted event (Connor, 1985).

8.6.8 Perception

Perception is greatly influenced by what the brain believes about objects and movements previously experienced in an environment, and what it expects the properties of those objects and movements to be (Rathus, 2008; Carter, 2009). Different components of the body's sensing and processing systems are involved when an individual perceives, processes, and responds/acts to a stimulus (Hawkins, 1997; Carter, 2009; Marieb and Hoehn, 2010).

Consequently, memories of previous perceptual experiences, such as practicing scenarios during simulation, play a strong role in influencing a person's perception, especially when cognitive resources are limited, and the results are seen in the user(s) behavior(s). Training is important because practice affords people with the ability to act in natural, intuitive, and instinctual ways. Such training increases the probability of a positive outcome when an abnormal situation expresses itself (Connor, 1985; Hollnagel, Woods, and Leveson, 2010; Hollnagel, Pariés, and Woods, 2011; Leveson, 2011).

8.7 ORGANIZATIONAL STRUCTURES (CULTURES)

Organizational structures, or cultures, influence human factors. The inter-relationships between management, regulatory leadership and operators can positively or negatively inspire the attitudes, motivation and overall behavior of the people within the organization (House et al., 2004; Sheehan, 2003; Lumpé, 2008; Rathus, 2008). Organizational structures are predisposed by national, societal, organizational, and professional cultures.

8.7.1 Safety culture

To provide enhanced risk mitigation, there needs to be an organizational structure that incorporates a just and safe culture (Salas and Maurino, 2010). A just and safe culture starts with the top executives and transitions through the appropriate leadership down to the operators.

I was very fortunate early in my career because the first airline I acted as a flight crew member for was Pacific Southwest Airlines or PSA. PSA incorporated a just and safe culture. There existed a significant difference in the cultures between PSA and the airline it was absorbed into. Unfortunately, those differences resulted in five fatal accidents in the five years post-merger. This demonstrated, to me, the importance of a robust corporate safety culture and how risk is increased when that culture is undervalued.

8.7.2 Management leadership style

The team-oriented leadership style is considered to be the superior leadership style (Lumpé, 2008). This leadership style starts with upper management and continues through the management hierarchy to the individual employee. The underlying forces of a team-oriented leadership style enhances (Lumpé, 2008):

- Company performance
- Customer service
- Customer loyalty
- Quality assurance
- Low sick time
- Increased revenue
- Lower costs (fixed and variable)

In a team-oriented culture, the relationships between management and employees stimulate an environment forming a team-oriented leadership structure, one that is "akin to family". This was the case at PSA.

Everyone felt that going to work was like going home and most employees could not wait to get back to work after the end of shift. Additionally, this type of culture enhances corporation benefits that further enhance safety such as:

- Increased productivity
- Increased quality assurance
- Increased employee longevity
- Increased passenger satisfaction

When the basic human needs of affiliation, belongingness, and love are met through the dynamics of a strong organizational culture, the resulting efficiencies of the organization create a highly proficient and safe company, lowering risk (Sheehan, 2003; Lumpé, 2008).

This type of leadership organization not only enhances stability for the company in long-term revenue gains, but also increases safety through increasing efficiencies in teamwork that are associated with the positive culture and, ultimately, lowers employee stress and mitigates risk and errors, as well as reducing incident and accident rates (Sheehan, 2003; Lumpé, 2008; Hamilton, 2011).

When a company incorporates SOPs that promote coordination and specific authority, responsibility, and accountability, the organizational culture influences the professional culture and therefore directly influences the probability of a positive outcome in the event of threats and errors; this reduces incidents and accidents (Sheehan, 2003; Lumpé, 2008; Hamilton, 2011). Finally, upper management should evoke the same dynamics to their team leaders who would be best served through the emulation of this type of culture that, according to Tullo, positively influences employees throughout the company (Tullo, 2010).

8.8 CONCLUSION

Risk in aviation is something that is associated with each flight and as the number of flights increase, there is an increased probability of the risk involved, inherently leading to an accident or incident. While a significant number of factors have been incorporated to offset those risks, as the world's number of daily flights increase; and the current, aging, professional pilots retire, and, if the current accident rate continues; there will be a significant increase in the number of related accidents per year.

Therefore, it is paramount that the regulatory entities, manufacturers, airlines, and unions continue to utilize the aspects of the above-described risk mitigation practices to continue to offset the increased risk that is mathematically associated with the increased number of flights and the increased number of new pilots who will be lacking the education and experience of the professionals that the industry now enjoys. "We know how to make the system even safer than it is, but we're going to lose ground if we fail to manage the growth within the limits of our human resources", says Bill Voss, president of the Flight Safety Foundation (Salas and Maurino, 2010, p. 294).

It is important to emphasize that designers from different domains and expert users need to work cooperatively when designing complex systems and developing methods to reduce the associated risks. User requirements must be well thought out from the beginning of the design phase and continue through the development and operational phases. This process allows designers to consider potential emergent properties that need to be addressed and solved. Also, because it is difficult and sometimes almost impossible to predict all conceivable emergent properties, scenario-based design, modeling and simulation, and HITLS need to be utilized. Using scenarios optimizes the dynamic interactions between the humans, the machines, and the human-machine system. It is during the HITLS that unplanned-for events will express themselves. This iterative process allows for the enhancement of procedural changes that will augment risk mitigation practices that supplement the operational cultures, policies, and environment; further reducing risk and enhancing safety, efficiencies, and comfort.

ACKNOWLEDGMENTS

I wish to thank the following individuals for their influence and editing skills: CA Bill Connor, ATP, Ph.D., U.S. Marines, Delta Airlines, and SETP (Society of Experimental Test Pilots) Associate Fellow; CA Pete Dunn, ATP, MS, U.S. Air Force, United Airlines, Florida Institute of Technology Visiting Professor; CA Greg Fox, ATP, Ph.D., Royal Canadian Air Force, CEO of CASSOS (Caribbean Aviation Safety and Security Oversite System), Florida Institute of Technology Associate Professor; CA Pat Murphy, ATP, Ph.D., Boeing Senior Test Pilot. Special acknowledgement to Coach Ron Vavra, MS, Grossmont College Track and Cross-Country Coach, for his continual support and positive influence for over 40 years. I would be lost in space without him.

REFERENCES

Baldwin, C.L. (2012). Auditory cognition and human performance. CRC Press. Boca Raton, FL.

Boeing (2012). Statistical summary of airplane accidents worldwide operations 1959–2011. Boeing Commercial Airplanes. Seattle, WA.

Boy, G.A. Ed. (2011). The handbook of human-machine interaction: A human-centered design approach. Ashgate. Burlington, VT. doi: https://doi.org/10.1201/9781315557380. 2017 eBook ISBN 9781315557380

Carter, R. (2009). The human brain book. DK Books. New York, NY.

Chappelow, J.W. (2006). Error and accidents. In D.J. Rainford and D.P. Gradwell (Eds.), Ernsting's aviation medicine, pp. 273–292. 4th Edition. Edward Arnold Ltd. London, UK.

Connor, C.W. (1985). Human performance capabilities – What are the operational capabilities. Technical report. SAE Aerotech. Long Beach, CA. October 14–17.

Curtis, M. & Jentsch, F. (2010). Line operations simulation development tools. In B. Kanki, R. Helmerich, and J Anca (Eds.), Crew resource management, pp. 399–420. 2nd Edition. Elsevier. San Diego, CA.

Cusick, S., Cortés, A. & Rodrigues C. (2017). Commercial aviation safety. McGraw Hill. New York, NY.

Dismukes, R.K., Benjamin, A.B. & Loukopoulous, L.D. (2007). The limits of expertise: Rethinking pilot error and the causes of airline accidents. Ashgate. Burlington, VT.

Dror, I.E. (2011). The paradox of human expertise: Why experts get it wrong. In Kapur, N. (Ed.), The paradoxical brain, pp. 177–188. Cambridge Press. Cambridge, UK.

Ericsson, K.A., Krampe, R.T. & Tesch-Romer, C. (1993). The role of deliberate practice in the acquisition of expert performance. *Psychological Review* 100, pp. 363–406.

FAA (2010). Aeronautical decision making. US Government. Washington, DC.

FAA risk management handbook (2016). US Government. Washington, DC.

FAA. FAR 121.403 (2017). Pilot training curriculum. US Government. Washington, DC.

FAA. FAR 121.407 (2017). Pilot training programs. US Government. Washington, DC.

FAA. FAR 121.409 (2017). Pilot training courses. US Government. Washington, DC.

FAA. FAR 121.433 (2017). Pilot training required. US Government. Washington, DC.

FAA. FAR 121.434 (2017). Pilot operating experience. US Government. Washington, DC.

FAA. FAR 121.437 (2017). Pilot qualifications: Certificates required. US Government. Washington, DC.

FAA. FAR 121.439 (2017). Pilot qualifications: Recent experience. US Government. Washington, DC.

FAA. FAR 121.443 (2017). Pilot in command qualifications. US Government. Washington, DC.

FAA. KLM Flight 4805 (2020). Boeing 747-206B, PH-BUF, ground collision with Pan American Flight 1736, Boeing 747-121, N735PA. US Government. Washington, DC. Retrieved from: https://lessonslearned.faa.gov/ll_main.cfm?TabID=1&LLID=52&LLTypeID=0.

Gawron, V. (2004). Psychological factors expectation. In F.H. Previc and W.R. Ercoline (Eds.), Spatial disorientation in aviation, pp. 145–195. American Institute of Aeronautics & Astronautics. Reston, VA.

Hamilton, J.S. (2011). Practical aviation law. Aviation Supplies & Academics, Inc. New Castle, WA.

Hawkins, F.H. (1997). Human factors. 2nd Edition. Ashgate. Hants, UK.

Helmreich, R.L., Kline, J.R. & Wilhelm, J.A. (1999). Models of threat, error, and CRM in flight operations. The University of Texas at Austin Department of Psychology. Austin, TX.

Helmreich, R.L., Merritt, A.C. & Wilhelm, J.A. (1999). The evolution of crew resource management training in commercial aviation. The University of Texas at Austin Department of Psychology. Austin, TX.

Hollnagel, E., Woods, D. & Leveson, N. (2010). Resilience engineering: Concepts and precepts. Ashgate. Burlington, VT.

Hollnagel, E., Pariés, J., Woods, D.D. & Wreathall, J. (2011). Resilience engineering in practice – A guidebook. Ashgate. Burlington, VT.

House, R.J., Hanges, P.J., Javidan, M., Dorfman, P.W. & Gupta, V. (2004). Culture, leadership, and organizations: The GLOBE study of 62 societies. Sage Publications. Thousand Oaks, CA. Retrieved from: https://www. andrews.edu/ services/jacl/articlearchive/1_1_summer_2006/6_br_globe_ (2004).pdf.

Hughes, R., Ginnett, R. & Curphy, G. (2011). Leadership: Enhancing the lessons of experience. 7th Edition. McGraw-Hill. Columbus, OH.

Jensen, R.S. (1995). Pilot judgement and crew resource management. Ashgate. Burlington, VT.

Kanki, B., Helmerich, R. & Anca, J. (2010). Crew resource management. 2nd Edition. Academic. San Diego, CA.

Kiss, D.M. (2011). Risk management in flight operations. Liberty University. Lynchburg, VA.

Kiss, D.M. & Stephane, A.L. (2016). Psychosocial and sociotechnical perspectives of risk. Florida Institute of Technology. Melbourne, FL.

Leveson, N.G. (2011). Engineering a safer world: Systems thinking applied to safety. MIT Press. Cambridge, MS.

Logan, G.D. (1985). Skill and automaticity: Relations, implications, and future directions. Canadian Journal of Psychology 39 (2), pp. 367–386.

Lumpé, M.C. (2008). Leadership and organization in the aviation industry. Ashgate, Burlington, VT.

Marieb, E.N. & Hoehn, K. (2010). Human anatomy and physiology. 8th Edition. Pearson. San Francisco, CA.

Nokes, T.J. Schunn, C.D. & Chi, M.T.H. (2010). Problem solving and human expertise. Elsevier. Philadelphia, PA.

NTSB (1972). Aircraft accident report: Eastern Airlines, Inc., L-1011, N320EA, Miami, Florida. U.S. Government. Washington, DC.

NTSB (1979). Aircraft accident report: United Airlines, Inc., McDonnell-Douglas DC-8-61, N8082U, Portland, Oregon. US Government. Washington, DC.

NTSB (1997). Aircraft accident report: ValuJet Airlines flight 592. US Government. Washington, DC.

Rathus, S.A. (2008). Psychology: Concepts and connections. 9th Edition. Wadsworth. Belmont, CA.

Salas, E., Bowers, C.A. & Edens, E. (2001). Improving teamwork in organizations. Lawrence Erlbaum Associates. Mahwah, NJ.

Salas, E. & Maurino, D. (2010). Human factors in aviation. 2nd Edition. Academic. Burlington, MA.

Sheehan, J. (2003). Business and corporate aviation management. McGraw Hill. New York, NY.

Sjöberg, J. (2000). Factors in risk perception. Stockholm School of Economics. Stockholm, Sweden.

Sumwalt, R.L. & Lemos, K.A. (2010). The accident investigator's perspective. In B. Kanki, R. Helmerich, and J Anca (Eds.), Crew resource management, pp. 399–420. 2nd Edition. Elsevier. San Diego, CA.

Telfer, R.A. (1993). Aviation instruction and training. Ashgate. Brookfield, VT.

Telfer, R.A. & Biggs, J. (1988). The psychology of flight training. Iowa State University Press. Ames, IA.

Telfer, R.A. & Moore, P.J. (1997). Aviation training: Learners, instruction, and organization. Ashgate. Brookfield, VT.

Tullo, F.J. (2010). Teamwork and organizational factors. In B. Kanki, R. Helmerich, and J Anca (Eds.), Crew resource management (pp. 59–78). 2nd Edition. Elsevier. San Diego, CA.

Wood, R.H. (2003). Aviation safety programs: A management handbook. 3rd Edition. Jeppesen Sanderson. Englewood, CO.

Chapter 9

Aging and risk-taking

An increasing dimension of life-critical systems

Anabela Simões[1]
Lusofona University
Lisbon, Portugal

CONTENTS

9.1 INTRODUCTION

When a system is complex and safety-critical, the more human opera-tors' skills, knowledge, competencies, and abilities are important for the system's efficiency and overall safety. However, the length of active life (around 40 years) means that human operators will age, and working situations and systems will evolve. Furthermore, it will not be possible that, in the 21st century, people spend their whole active life performing the same kind of work using the same operational methods or tools, as was

DOI: 10.1201/9781003221609-9

the case in most working settings during the last century. Nowadays, rapid advances in technology foster changes in working and everyday utility systems. Thus, variability (as opposed to stability), uncertainty (as opposed to certainty), dynamics (as opposed to static), and the unexpected (as opposed to the expected) are elements of the increasing complexity of nowadays systems that one has to deal with. This is the new norm for current daily life, which requires a strong educational background stimulating creativity and critical thinking as the main resources to adapt to innovation and face the unexpected. When aging is added to this scenario, special attention must be paid to managing operators' age and function allocation. As stated by Boy (2020), *"becoming familiar with complexity is often much better than simplifying complexity"*. Thus, developing specific skills, accumulating experience on the job, and becoming familiar with complexity are the key elements for long-term success on the job.

Despite age-related declines, older workers develop a process expressed by a capacity to compensate for such declines, which is supported by individual skills, knowledge, and experience on the job, as well as the expertise they have developed stimulating flexibility and adaptability. It could be said that the best conditions for a continuous development of skills and knowledge, together with increasing experience on the job throughout active life, are required to stimulate such self-regulatory processes. This was easy when people used to perform their jobs using the same tools and processes for many years. Nowadays, the continuous and rapid development of technology leads to frequent changes at work, which requires an increased ability to adapt, together with creativity and critical thinking, to stimulate the ability to compensate and ensure the best performance on the job throughout working life. Therefore, older operators will be able to deal with unexpected and emergency situations requiring new solutions to overcome any crisis. Age-related declines are not that evident during active life due to this compensation capacity, and the significant cognitive stimulation represented by the work-related demands slowing age-related declines. Thus, age-related declines are much more evident after retirement due to the lack of daily cognitive stimulation represented by work demands.

Another aspect related to aging that should be highlighted is a greater sensitivity to transient factors (fatigue, distraction, health conditions, psychological state, etc.), which can temporarily compromise Situation Awareness (SA) and impair an operator's ability to respond to any emergent request. Within life-critical systems, professionals have been highly trained, are used to taking risks, overcoming them, and surviving throughout their professional life. These conditions lead them to build up the required experience and ability to handle the situation at hand or event-related uncertainty and take the necessary risks. Such experience, as well as the related knowledge, skills, and special competencies, will last for as long as they are used, although with increasing age, everyone will need to develop his or her own compensatory strategies to overcome some temporary difficulties in each specific situation. Although common people tend to be more conservative and appreciate stability with increasing age, these professionals are highly competent, skilled, and prepared to handle uncertainty and take the necessary risks.

9.2 DEMOGRAPHIC TRENDS

According to the United Nations (2004), the elderly population in the world is growing at its fastest rate ever. It is expected that by 2050 the world will be home to more than 2 billion people aged 60 or over. In the USA, the number of people aged 65 years and older is projected to grow from about 35 million in 2000 to more than 86 million in 2050, increasing the population by 20.2%. Data from the Census Bureau dated December 2010 show that within the next four decades most developed countries in Europe and East Asia will become old-age homes, with a third or more of their populations aged over 65.

Every demographic projection for the most industrialized countries predicts a significant increase in the number of people aged 65 years old and older. The so-called graying of Europe refers to the increase in the proportion of elderly people in the population compared to the workforce. This is a social phenomenon resulting, on the one hand, from the *baby boom* generation having now reached their early seventies, and, on the other hand, from a decrease in birth rates and an increase in life expectancy at birth. Aging of the global population is considered "... an unprecedented, new phenomenon, all pervasive, profound and an enduring change in nature" (UN in Ilmarinen, 2005). As older people represent the most rapidly growing segment of the population in all developed countries, this phenomenon has led to significant changes in the age structure, and it is also creating new challenges for society, companies, and individuals themselves.

As an indicator of well-being, life expectancy is translated onto an increase in the expected number of healthy and functionally unrestricted years. This means that older people will be able to carry out an independent life until a greater age if the current mortality and morbidity trends remain unchanged (Ilmarinen, 2005). As a natural consequence, working life is being extended in line with these demographic trends, which raises new questions: how to work in later life dealing with complexity, uncertainty, the unexpected, and risk exposure when facing age-related decline? Will COVID-19, with the high number of deaths around the world affecting mainly the elderly population, change these demographic trends? Currently, the future of the coronavirus on the planet, taking into account mutations occurring over time, is unknown. The above-mentioned demographic projections are therefore very uncertain. Consequently, working toward resilience will empower countries, cities, sociotechnical systems, and people to prevent and fight every kind of attack by making the necessary adaptations to survive and recover.

9.3 AGING THROUGHOUT WORKING LIFE

With increasing age, individuals experience some level of functional decline, resulting from changes in sensation, perception, cognition, as well as psycho-motor and motor skills. These declines are reflected in difficulties in discrimi-nating the relevant information and decision-making, particularly in dynamic and complex environments and mainly under time pressure. Furthermore, they may induce some limitations in dealing with uncertainty and unexpected events. Because of individual differences in aging, there is considerable vari-ability among older people in terms of physical, psychological, and behavioral changes with increasing age. These difficulties express their needs for more time to select and process the available information toward the appropriate decision-making (Eby et al. 2000, 2009; Simoes, 2003). Age-related declines may also induce some limitations in promptness in reacting to critical situa-tions, particularly regarding safety and security issues.

Aging is an important dimension in every working system and should be managed before identifying any signs of deterioration in performance that could potentially lead to an increase in the risk of major incidents or acci-dents. This identification requires the development of strategies to prepare for and accommodate an aging workforce. According to Czaja and Moen (2003), to do so involves understanding (1) the characteristics of older workers and the growing population of older adults who no longer work; (2) the potential implications of aging at work and work environments; (3) the technological and social characteristics of existing jobs and work environments; and (4) the triggers, dynamics, and processes moving people into and out of work. Therefore, it will be impossible to maintain previ-ous attitudes whereby older workers were excluded. Life-span education and training policies must ensure their adaptability to technology-related changes to life and work contexts and conditions by providing appropri-ate training addressing the development of adaptability and creativity as workers age. Thus, two important aspects must be considered: (1) learning and acquisition of new skills should be based on previous knowledge and experience, which is the prime request for the success of learning and will eliminate any feeling of anxiety or resistance to change; and (2) a human-system integration approach should be applied to ensure the system is user-friendly, and thus allows for easy and efficient human-system cooperation.

Unlike machines, humans evolve over time, learning, acquiring experi-ence, aging, and being temporarily tired or sick. Due to distinct internal and external influences, each individual will age differently and, for the same person, each biological function will age at a different pace. This results in a great variability in terms of age-related physical, psychologi-cal, and behavioral changes, as well as differences between subjects. Even healthy older adults are affected by the age-related decline, leading them to experience some difficulties in performing specific tasks imposing sig-nificant demands and, particularly, in dealing with decision-making under stress or time pressure. Thus, the duration of active life means that human operators will age, and some abilities will decline – although workers can stay competent thanks to their experience performing the task in the context at hand. Therefore, the more skilled and competent operators are, the more efficient the self-regulation process will be, allowing for: (1) the best compensation for their functional declines, staying highly competent and thus, coping with the increasing complexity and dynamics of sociotechnical systems; and (2) a reduced influence of transient factors (fatigue, distraction, fluctuating health state, etc.) on their performance.

9.3.1 Aging and self-regulation behavior

Despite age-related declines, older people perform their tasks safely and efficiently, compensating for their failures by using their remaining abilities more fully. For some common tasks, it seems that, for the same task performance, the same types of compensations for functional losses can be found among older people, resulting in common patterns that are different from those observed in younger people (Eby et al., 2009). Therefore, older people compensate for their impaired sensorimotor functioning by adapting their behavior according to the circumstances in which the task is performed and by using their still available compensatory potential (Czaja & Moen, 2003). As the ability to compensate for age-related declines is related to individual experience in each specific task, the working context could be easier for older workers due to their longer experience on the job. Thus, a greater experience of the task and context should lead to optimized performance, which becomes more regular, precise, and rapidly executed, less effortful and more automatic, and less influenced by transient factors (Figure 9.1). This should result in an increased ability to anticipate potential risky situations at an early stage from a few slight cues, and consequently to react in an adequate way and in due time when required. This increased control may improve the potential to rectify errors, provided enough time is available.

This self-regulation process, expressed as a capacity to compensate for age-related declines, is supported by each individual's skills, knowledge and experience on the job, provided the following conditions are met: (1) abilities are matched to job requirements and expertise; (2) working conditions (technical, organizational, and environmental) are appropriately designed to avoid unnecessary adaptation efforts; (3) experience on the job can be mobilized to compensate for some decline in functional abilities; and (4) lifelong training is provided aiming to maintain and improve key skills and stimulate creativity. However, the current increase in system complexity requires the following key elements for experience to be mobilized: familiarity with complexity, and maturity of practice.

Self-regulation behavior has been studied in older drivers. Eby, Molnar & Kartje (2009) classify self-regulation behaviors at three distinct levels: strategic, tactical, and life-goal. Strategic self-regulation refers to decisions

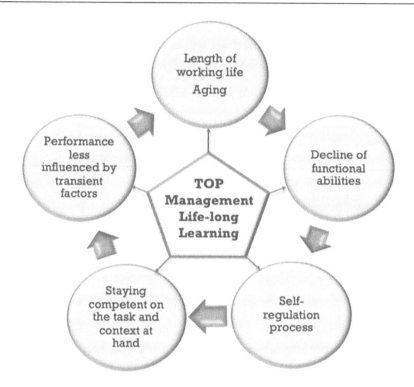

Figure 9.1 The cycle of influences leading to the control of performance variability.

made by drivers before starting a driving trip, avoiding the busiest times or roads, driving at night, or if it is raining. Tactical self-regulation refers to being focused on the driving task and avoiding distractions, such as secondary tasks. The life-goal self-regulation behavior refers to decisions about where to live in relation to frequent destinations or what type of car to drive with the aim of facilitating the driving task and avoiding risks. These decisions allow older drivers to continue driving and managing the task according to their abilities and avoiding conditions that could make it difficult or risky. Usually, older drivers prefer a very conventional car without recent technology, the use of which can become a factor of distraction. Thus, the most relevant factors to build a self-regulation behavior are experienced on the task as they have known it for years. Within a professional context, skilled, experienced, and expert professionals develop an effective self-regulation behavior to be applied, when necessary, in complex and risky situations requiring their compensation potential to perform the task as planned and survive.

Within life-critical systems, some decreases in performance could be disastrous. One important factor to allow for a successful self-regulation behavior is risk-awareness, which, at a professional level, is usually encouraged through specific programs. Hopkins (in Borys, 2009), defines risk-awareness as a cultural approach to safety, arguing that *"a culture of risk-awareness is interchangeable with the notions of an informed culture and collective mindfulness"*. According to the author, risk-awareness is enacted through organizational practices rather than by trying to change the mindsets of individual workers. However, rather than identifying those contexts, it would be better to explore the specific tasks within a job and how work is organized in order to identify life-critical implications resulting from decreased performance. Decision-making speed has been reported to decline with increasing age, particularly in unfamiliar conditions. Thus, in highly demanding situations, individuals may be unable to use their experience or expertise, which compromises decision-making. Jobs where individuals may find themselves in unusual conditions and/or unfamiliar environments could result in a decrease in performance and, consequently, in undesired outcomes.

9.3.2 Aging and maturity of practice

Many critical situations require immediate action before complete information becomes available about the conditions. This requires decision-making based on extensive knowledge, skills, experience, courage, and creativity to achieve a survival performance level instead of a commonly targeted performance level. Considering the system as a whole as a joint cognitive system, the human-technology cooperation aspect should be viewed as a cooperative relationship based on a match between the human and the machine. In this case, the technology assists the user in performing the task by means of functional cooperation between the operator and the technology. An assistive device can also be considered as support technology, helping older workers by easing their compensatory behavior. This perfect coupling between humans and technology requires good usability resulting from a human-system integration approach during its design and development.

Highly skilled, competent, and trained professionals, as required in life-critical systems, accumulate experience and improve their skills and competencies as they age; on the technology side, the diversity of tools they have to deal with when performing tasks, together with the related rapid evolution, may lead to an increase in complexity and uncertainty. To avoid difficulties in dealing with the technology resulting from poor usability, a human-system integration approach should be favored during the technology design and development (Boy, 2020). Targeting safe, efficient, and comfortable interactions with technology to allow the completion of their goals, operators should appropriate technology as an extension of their own functioning and abilities, building their adaptation and mental representation as the basis for their trust in the technology. This attitude requires them to view the reliability and robustness of the technology as qualities. Then, they take risks, taking into account the required information and possibly by anticipating the situation they will find, based on their familiarity with complexity, their skills and experience. This successful appropriation of technology according to criteria of safety, efficiency, and comfort represent the **maturity of practice** (Boy, 2013, 2020).

Before the maturity of practice, the evolution of technology toward a maturity level is a main request for system efficiency and safety. The maturity of technology, usually expressed as a Technology Readiness Level (TRL) (Boy, 2013), indicates when the technology is ready to be used. However, it is necessary to provide information and training together with the management of possible resistance from potential users. Furthermore, it is expected that a technology will evolve during its life cycle. Therefore, both technology maturity and practice maturity must be managed so that the available technologies will remain reliable, robust, and flexible, and to ensure that human-technology interactions and cooperation will be efficient, safe, and comfortable.

9.4 AGING, VARIABILITY, AND UNCERTAINTY

Task performance within complex systems imposes safety and security requirements that depend on each specific context. Much of fault tolerance relies on the system characteristics creating paths to a safe performance or to human error. To err is human, and this cannot be changed, but the conditions leading to erratic behavior can be changed by adapting organizational and environmental conditions to the task and to human functioning. Human operators within a system not only represent the most valuable and reliable elements but also the most vulnerable. This vulnerability is due to human variability and instability (diversity, state of health, fatigue, etc.), as well as human behavior, which together with some external factors and organizational constraints can influence how the task is performed.

Human variability is defined as the range of possible values for any measurable human characteristic. Differences can be trivial or important, transient or permanent, voluntary or involuntary, congenital or acquired, and genetic or environmental. This means that everyone is distinct due to biological inheritance, education, cultural, and social environment, which is known as interindividual variability. Individuals are also subject to within-subject variability due to transient factors (fatigue, health condition, consumption of alcohol, drugs or medication, emotional state, stress, etc.), as well as aging. Age-related declines during active life increase performance variability, but the experience accumulated in performing the task and related to the context will increase the operator's competence, particularly when dealing with complexity, system failure, or any emergency situations. As people are so different from each other, and are also subject to internal variability, this can lead to significant dispersion of performance even if the circumstances are rigorously identical. However, among highly trained and skilled groups of professionals, it seems that age-related decline generates the same types of self-regulation behavior, which reduces performance variability among older workers. Thus, the variability of human performance, rather than being viewed as a constraint, should be viewed as the potential ability to recognize, adapt to and absorb variations, changes, disturbances, disruptions, and surprises, especially disruptions falling outside the set of perturbations the system is designed to handle.

A performance based on following rules, procedures, and instructions to the letter, which are usually incomplete, will be both inefficient and unsafe. To compensate for this incompleteness, individuals and organizations habitually adjust their performance in line with current demands, resources, and constraints, which is central successful performance; but since information, resources, and time are limited, particularly in dynamic and complex environments, any adjustments will inevitably be approximate. For this reason, performance variability is unavoidable and should be recognized as a source of success, even if it can also be a cause of failure. To avoid this and stimulate the operators' ability to adjust their performance to actual conditions and system variability, there is a need for a continuous improvement to their skills and competence, stressing their decision-making and anticipation abilities along with a strong stimulus for their creativity.

9.4.1 Transient factors of human variability

Transient factors are events that can temporarily modify the human's internal state; they include fatigue, drowsiness, distraction, health condition, etc., and may have a negative impact on task performance. These factors, combined with the related motivational processes and contextual factors, should be understood to be appropriately managed to ensure safety and efficiency. The influence of variability in health conditions on work ability is also a relevant aspect and is twofold: a good health condition has a positive influence and is an important factor in promoting work ability; a transient poor health condition has a negative impact on work ability and, consequently, on how the task is performed and on safety. Age-related declines and/or some medication can have a negative impact on the individual internal state and task performance. On the one hand, this can be due to age-related decreased rate of information processing or to any specific dysfunction; on the other hand, some prescribed drugs (tranquillizers and sedatives) can potentially adversely affect work ability and individual performance as they slow reaction time and diminish hazard awareness (Oxley et al., 2005; Simoes, Pereira & Panou, 2011).

9.4.1.1 Fatigue

Today society imposes continuous operation, which requires some systems to work around the clock in order to satisfy societal needs. This has resulted in continuous work, but the human operator is unable to remain continuously active throughout the day. Thus, the conflict between human capabilities and today's 24 h society leads to operator fatigue and stress, compromising performance levels, and potentially leading to accidents.

Fatigue is a short-term consequence of an activity that is expressed by some subjective symptoms and a decrease and instability in performance. Fatigue-related loss of control is one of the main factors leading to accidents. Most of the underlying causes are related to sleep deprivation resulting from circadian factors associated with work schedules and undiagnosed or untreated sleep disorders. The time spent on the task and the use of sedating medications or alcohol are also fatigue-causing factors, and a combination of any of them can greatly increase one's risk of fatigue-related failure.

A model of fatigue based on the sensorimotor capacity involved in the task performance for prolonged periods (Desmond & Hancock, 2001; Mathews et al., 2012) explains the occurrence of fatigue in several activities. This model distinguishes between active and passive fatigue, the first resulting from continuous and prolonged task-related perceptual-motor adjustment required when performing dynamic and effortful tasks requiring both physical and cognitive (attentional) efforts. Passive fatigue, in contrast, is related to situations where the person appears to do nothing for long periods, like driving long distances or verifying a dynamic process, particularly in monotonous situations or at night. Therefore, both situations lead to fatigue, but only passive fatigue leads to drowsiness as the low attentional demands decrease the level of vigilance. In addition, the stress related to both working conditions is considered a safety problem, as the statistical association between stress and accident risk is well established.

9.4.1.2 Attention

Attention is a cognitive function with two interdependent dimensions: *selectivity*, referring to the selection of the relevant stimuli coming in through our senses; and *intensity*, which is the dimension of alertness and thus refers to focusing on one area to deal with it effectively (Van Zomeren & Brouwer, 1994). When performing control tasks, particularly in a dynamic decision-making environment, continuous information processing is required to achieve the task purposes in terms of safety and efficiency. In such cases, attention represents the cognitive function that is the most required. A lack of attention resulting from passive fatigue and drowsiness, or high workload is often behind mismatches or even accidents. Attention overload may also be a problem for unstressed operators; worry and fatigue tend to impair functional attentional efficiency so that the stressed individual becomes particularly vulnerable to overload (Mathews & Desmond, 2001).

9.4.1.3 Distraction/inattention

Distraction is defined as a diversion of attention away from ongoing activities toward a competing activity, which is critical to a safe performance (Lee, Young & Regan, 2009). When distracted, the individual does not pay the necessary attention to the task and is thus unable to react appropriately in due time to critical events in the environment.

Although distraction is triggered by an external event or object, voluntary or not, inattention (Chapon, Gabaude & Fort, 2006) is not triggered by a specific event. Rather, it is a non-intentional endogenous distraction. Inattention occurs in routine situations that do not require sustained attention from the individual and thus, leave him/her free to get lost in his/her thoughts.

With increasing age, some declines in perceptual and cognitive functions affect task performance, particularly in complex, dynamic, and safety-critical environments (Holland, 2001). As a result of such declines in selective attention and attention switching, older adults have difficulty staying focused on primary information and preventing their attention from wandering to irrelevant material.

9.4.1.4 Situation awareness

Defined by Endsley (1995) as the perception of the elements in the environment within a volume of time and space, the comprehension of their meaning and the projection of their status soon, SA has a major place in decision-making, particularly in the presence of potential risks. This definition represents the continuous picture of the surrounding dynamic environment perceived and interpreted by an individual while performing his/her dynamic activity to guide each decision made toward achieving the pursued goals. Considering that this process is dynamic and evolves within a constantly changing environment, the sequence of images will guide any projection of such evolution and, thus, the required decision-making. However, a high level of task-related expertise, together with both knowledge and experience with the interactive system (maturity of practice) (Boy, 2016) and familiarity with the environment, will facilitate appropriate and effective decision-making, leading toward the completion of the pursued goals in safety.

In a study investigating age-related differences in the ability to perceive important abstract information in the environment, older adults demonstrated reduced SA compared to younger and middle-aged adults (Bolstad, 2001). The factors identified for this were related to their useful field of view, perceptual speed, and self-reported vision, but can be compensated by their experience performing a task (Caserta & Abrams, 2007) identified the following main cognitive declines in older adults: fluid intelligence, capacity for inhibition, number of processing resources, and speed of processing. However, recent research on cognitive training has shown that such declines can be contained to improve older adults' performance in several tasks (Caserta & Abrams, 2007; Simoes, 2003). Furthermore, such strategies to contain and recover from cognitive declines also have the potential to improve the person's age awareness and perception of his/her own progression, which are important to reduce the effects of cognitive aging (Fu, Kessels & Maes, 2020).

9.4.1.5 Physiological and mental condition

Some individual decisions relating to lifestyle, rest, and sleep habits, as well as the consumption of alcohol and other substances affecting task performance, particularly for complex tasks, are behavioral decisions that are rooted in basic attitudes underlying lifestyle. Within life-critical systems, education, training, and the existing regulations taken together lead to appropriate behavior on and off duty. Furthermore, the bodily function should be understood as being related to the activation level, which is the required physiological support for the task to be performed. It is a dynamic variable that is subject to continuous change during the individual activity and has a significant influence on workload. Targeting the balance between task demands and the available functional resources, an overload could result from high demands balanced with insufficient available functional support. Then, a decreased functional level, rather than resulting from a depletion of the required resources, could be related to a temporary difficulty in mobilizing the available resources to perform the task. These types of difficulty may result from a temporary impairment related to fatigue or sleep deprivation, which could be overcome through external stimulation or a voluntary attentional effort. Within life-critical systems, task performance imposes high levels of attention and promptness to react appropriately to unexpected situations in due time. Consequently, any behavior with the potential to decrease or impair the individual abilities required to ensure safe performance must be avoided.

As older people are more sensitive to variations in health conditions and related drivers, older professionals working in life-critical systems are subject to the abovementioned regulations and more frequent health checks, particularly to monitor biological functions with more direct influence on judgment and task performance. A periodic assessment of work ability, particularly addressing the age-related evolution of skills, competencies and motivations, should be an established practice within life-critical systems.

9.4.2 Managing variability and uncertainty

Managing variability and uncertainty within a life-critical system is a matter of enhancing competencies and skills so that people may become more adaptable to highly-paced changes in performance conditions and prepared to face the unexpected. Some pitfalls in this aim for high competencies and team homogeneity arise as an effect of transient factors together with interindividual differences.

According to Jackson (2010), the operation of life-critical systems tends to be highly unpredictable, as decisions and actions, once initiated, can rapidly produce chain reactions, and therefore become irreversible and difficult to trace. This aspect, together with human variability, introduces uncertainty into the system, which requires that human operators can cope with it. With increasing age, human operators become less comfortable with uncertainty, as they prefer more stable environments; however, highly skilled and experienced professionals used to deal with uncertainty across their professional life will continue being prepared to make decisions under extreme conditions based on their skills and experience.

When performing a task, the human operator mobilizes different cognitive functions to maintain his/her cognitive stability (Boy, 2003): anticipation (preparing the reaction based on the perceived information, existing knowledge, and experience), interaction (reactions to an external stimulus following a previous intention or plan), and recovery (conscious actions following a diagnosis of the situation). However, the automatic mode is more likely to be activated under conditions of stress and time pressure, but older workers are not comfortable acting under pressure and have greater difficulty dealing with decision-making in such conditions. Thus, operators who are more competent, skilled, and used to dealing with uncertainty are prepared to find solutions for any sudden problem, consequently older operators can better mobilize the acquired resources and make appropriate decisions in due time. Extensive system variability, together with some lack or ambiguity of information, particularly in dynamic environments, can lead to uncertainty. Age-related declines and a tendency to decrease risk-taking behavior with increasing age, introduce difficulties in decision-making in the face of uncertainty. Thus, highly skilled, competent, and experienced people have developed adaptability and maturity in their practice, making them autonomous enough and prompt to make decisions will be able to overcome difficulties and successfully maintain system stability (Figure 9.2). These individuals produce appropriate outputs by adopting different strategies when compared with younger and less experienced people.

Existing findings from research on aging and work suggest that due to some decline in cognitive resources, older workers tend to make more errors when performing their tasks (1) under stressful conditions resulting from time pressure or a high workload or (2) under substandard personal conditions resulting from sleep deprivation, fatigue, or a temporary health condition. Although younger adults are also sensitive to these situations, older people present lower performance thresholds, which can be exceeded at lower workload levels.

Research carried out on decision-making under pressure, which is recurrent within safety-critical systems, provides the following findings

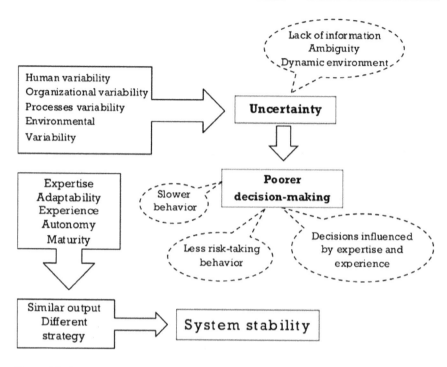

Figure 9.2 Overcoming age-related declines under uncertainty.

(Buljan & Shapira, 2005): (1) decision-making under time pressure leads to lower performance; (2) motivational pressure influences attention leading to slower information processing; (3) the degrees of freedom when managing more than one task simultaneously are severely constrained; and (4) it has also been demonstrated that deadlines have a negative effect on creativity. These findings suggest that when people operate under pressure, their attention is focused on a specific target and they cannot mobilize sufficient cognitive resources to follow several targets simultaneously. This can be explained by the way human beings control their actions, which involves different combinations of two control modes (Rasmussen, 1987): the conscious and the automatic. The conscious mode is slow, sequential, and logical, but limited in its capacity, being used to paying attention to something. The automatic mode is unconscious, which means that the individual performs an automated task while remaining aware of its evolution but not of the process controlling the succession of actions. It is very fast and allows different actions to be performed in parallel, being virtually unlimited in capacity. This mode is related to the operator's experience in performing the task, which is older operators' main resource as it allows for compensation behavior to reduce mental workload and stress.

9.5 A WORKING LIFE WITHIN LIFE-CRITICAL SYSTEMS

Safety, efficiency, and comfort are the three major dimensions of life-critical systems (Boy, 2013), but aging is an additional part of the puzzle that should be taken into account in system design and operation to avoid compromising any of those dimensions. Thus, skills and knowledge should be continuously developed with experience and lifelong training to stimulate adaptability to innovation, as well as the right risk-awareness and the required creativity to manage challenges and provide opportunities to adopt the appropriate decision.

The more developed and technology-based working systems are, the more safety-critical they become, thus imposing fast, efficient, and precise actions on the humans involved. Modern organizations are increasingly shaped by technology (automated processes, computer-based cooperation networks, multimedia applications, etc.) integrating social and organizational elements, which are essential to understand the system's behavior and should be viewed as embedded in the system (Tschiersch & Schael, 2003; Boy, 2020). When a system is complex and safety-critical, the more human operators' skills, competencies, and abilities are important to ensure high performance levels for efficient and safe functioning of the system. To ensure the success of operators' actions, resulting from good cooperation between humans and technology, the different factors leading to performance variability must be identified, understood, and managed. These factors are triggered by either internal or external factors impacting the task performance as a function of the diversity of human characteristics and functioning, as well as variability in the short-term and of life span. This contradicts the frequent assumption of stable human activity over time that characterized the design and management of numerous working systems for many years. Indeed, there is no such thing as the average human being; human variability, resulting from diversity or the instability of human activity, is an uncomfortable reality that systems designers and managers have to face and integrate accordingly. Although the management of modern companies is now increasingly finance-driven, placing money at the center of the decision-making process, the importance of the human operator's skills, knowledge, experiences, and creativity is recognized to ensure both safety and efficiency, particularly, within life-critical systems. Furthermore, modern and technology-based systems tend to apply the TOP model, which is a framework integrating Technology, Organization, and People since the initial design phases (Tschiersch & Schael, 2003; Boy, 2020). Currently, people represent the most flexible, adaptable, and valuable elements of any system. They learn with experience as adaptive, responsible, collaborative, and tool creating/wielding agents to achieve success. Even under resource and performance pressure at all levels of the system, they can learn and adapt to different situations and multiple task goals. They are also the most vulnerable elements in a system, due to human variability and instability, as well as human behavior, which, together with some external factors and organizational constraints, can influence human performance. Thus, the integration of human characteristics, behavior, and variability into system design since its early design phase (human-system integration) is required to improve the system's efficiency and overall safety (Boy, 2013, 2020). Targeting adaptation, potential users/operators must be considered in the design process. Boy (2016) proposes a set of criteria based on the TOP model for the assessment of human-system integration in the design of a life-critical system. These criteria consist of a set of shared, but not exhaustive, features for each element of the model:

- T (Technology) – robustness, reliability, resilience, efficiency, technology maturity, modeling and simulation, and advanced interaction *media*;
- O (Organization) – communication, cooperation, coordination, orchestration, social issues, environment, organizational complexity, organization design and management, and function allocation;
- P (People) – safety, training, occupational health, skills, knowledge, performance support, and life criticality.

Life-critical systems are operated by humans and are thus exposed to human-related disturbances resulting from several interlinking contextual factors. The nature and dimensions of any disturbance depend on the task being performed, the individual's skill level and their functional abilities, and state, as well as the local conditions for system performance and some related organizational factors (Reason & Hobbs, 2003). The prevention of unwanted and unplanned disturbances resulting directly from unsafe behavior is the main problem; therefore, conditions for the success within a life-critical system should represent the central point in its design and operation. Considering the whole system as a joint cognitive system, human-technology cooperation should be viewed as the state of being in action within a system in a perfect coupling between the human and the technology, which is the basis for the system's efficiency and safety. The great human variability is a source of uncertainty, but it is also related to the ability of human beings to constantly adjust their performance using their competence, experience, and creativity with the aim of maintaining a certain degree of stability. Together with transient factors of variability creating uncertainty, aging, and the related effects on functional abilities represent an important concern that should be managed.

9.5.1 Aging and risk-taking behavior

Risk is part of our life since birth. Growing up and personal development are a succession of overcoming risks until standing, walking, running, and developing the skills required for an autonomous life. Every action, particularly actions having uncertain outcomes, requires risk-taking, which is an inevitable behavior in any complex and dynamic environment.

Studies addressing risk-taking can be found in the literature but, on the one hand, they cover different domains like financial, social, ethical, health, recreational, or gambling; on the other hand, studies focusing on age differences involve samples of older and retired people. One study on risk-taking differences across the adult life span (Rolison, Hanoch, Wood & Liu, 2013) is a cross-sectional study involving 528 participants aged from 18 to 93 years. This study addressed five of the abovementioned domains (ethical, financial, health, recreational, and social). The age range of the older group (N = 107) was 52–93 years old, with representatives of both genders. Previous studies in the field of Psychology identified older adults as risk avoidants stating that risk-taking behavior decreases with age across domains (Rolison, Hanoch, Wood & Liu, 2013). However, the findings of this study indicated that age-related differences in risk-taking are specific to the domain. Risk-taking in some domains (social and recreational) decreases more abruptly from young to middle age than in the older group. Health and ethical risk-taking reduce progressively with increasing age; risk-taking in the financial domain reduces steeply in later life. Another study on risk-taking compared young and older adults in a laboratory task (the Columbia Card Task) (Huang, Wood, Berger & Hanoch, 2013). No differences were found in the performance of the two groups, but the authors concluded that a laboratory-based study was not appropriate to analyze risk-taking behavior in older people as they may act differently in the real world where they may rely on others in their decision-making.

Several studies have compared frequent and occasional risk-taking behavior between different age groups, including adolescents and younger adults, focusing on the related triggers. It seems that sensation seeking, age, and negative affectivity modulate frequent risk-taking behavior, whereas just sensation seeking contributes to occasional risk-taking behavior. Furthermore, simple parental control or a lack of opportunity can prevent younger people from engaging in frequent risky behaviors. They seem to be more susceptible to the influence of their peers in risky situations, which means that adolescents and younger adults take risks rather for pleasure and to meet a need for acceptance by their peers rather than to deal with a real need to overcome a difficulty or for survival. Therefore, they take risks for no reason in a total absence of: (1) any assessment of the conditions they will find in relation to their available resources, (2) any knowledge they have with which to understand the situation and direct the appropriate decision, or (3) the ability to anticipate how the situation will evolve. These are negative risk-taking behaviors. With increasing age and education, this risk-taking behavior will change, and the same individuals will adopt more responsible attitudes leading to appropriate behavior in risky situations. At this stage, risk-taking behavior in particular situations will be led by context-related knowledge, awareness of existing risks, the equilibrium between the needs of the situation and the resources available to make the appropriate decision and perform the related action successfully. This is positive risk-taking behavior provided it is conscientious, knowledge-directed, and supported by the required skills and competence – i.e., knowing what to do.

9.5.2 Risk-taking, uncertainty, and decision-making

Taking a risk means that a decision must be made and the corresponding action should be performed, sometimes in the absence of complete information about the conditions that will be found and/or the related outcomes. Furthermore, decision-making in real-world settings, particularly in complex and dynamic environments, involves an important relation between time and action, as the right action must be performed in due time; otherwise, the same action may no longer be suitable. According to Boy (2013), decision-making and risk-taking are closely related: (1) both involve important and related cognitive processes; (2) decision-making entails risk-taking; (3) risk-taking involves action resulting from a decision-making process with more or less knowledge of its possible outcomes; and (4) risk-taking requires preparation, training, knowledge, and discipline, together with SA, which is an important part of the regulatory loop in human activity (Bellet et al., 2011). Every action, particularly those with uncertain outcomes, requires risk-taking, which is an inevitable behavior in any complex and dynamic environment, and it is necessary when there is no prescribed solution to deal with a particular situation.

Any decision in real-world settings and its outcomes are associated with a degree of uncertainty, as risk is an inherent part of everyday life and is present in everyday decision-making. As stated by Vertzberger (1998), the risk is a "real-life construct of human behavior representing a complex interface among a particular set of behaviors and outcome expectations in a particular environmental context". Risk is therefore associated with uncertainty, particularly regarding the outcome value (in terms of being positive or negative, desirable or undesirable) and ambiguity (in terms of being known or unknown). The level of risk is another issue that is defined by the answers to the following questions: (1) What are the gains and losses associated with each known outcome? (2) What is the probability of each outcome? (3) How valid are the outcome probabilities and gain-loss estimates? Thus, the risk is the "likelihood that validly predictable direct or indirect consequences with potential adverse values will materialize, arising from particular events, self-behavior, environmental constraints, or the reaction of an opponent or third party" (Vertzberger, 1998). From this perspective, risks can be estimated according to: (1) the desired or undesired outcome values, (2) the probability of outcomes, and (3) the validity attributed to the estimates of outcome values and probabilities.

In the absence of complete information about the conditions people will find, decision-making is based on the maturity of practice combined with courage and creativity to achieve a survival performance level rather than a commonly targeted performance level. This is the case for safety and life-critical systems where fault tolerance absorbs variability in human performance and thus becomes a way to enhance the system's reliability. Furthermore, taking a risk is not an isolated decision once each potential risk-taker is a member of a team and the success or failure of his/her action will positively or negatively affect the team and, probably, the whole system. This requires professionals with high levels of skills, competency, and expertise, being prepared: (1) to take risks out of any predefined solution to deal with potential problems, and (2) to deal with the unexpected. Thus, risk-taking requires training, knowledge, skills, and experience to mobilize the required attention to focus choice on success and survival (Buljan & Saphira, 2005). Depending on the operator's own perception of their abilities and skills, as well as their maturity of practice, there will be a decision directed by an attitude that can vary from a heightened awareness of the situation to high-variance alternatives, which will lead to risk-prone behavior. Therefore, risk-taking behavior depends on: (1) the available resources and related self-perception; (2) risk-taking perceived as success or failure; and (3) the way attention is allocated between both reference points (targeted performance or survival) (Buljan & Saphira, 2005).

9.5.3 Aging and risk-taking

A few studies into the effects of age on risk-taking address older adults and their risk-taking behavior but they are laboratory studies offering a very limited environment compared to the real world. Furthermore, every study involving participants of older ages either addresses retired people, or a particular field in life, or a comparison between different age groups, genders, and fields. So far, research carried out on risk-taking and age does not address aging professionals from safety-critical or life-critical systems. Some studies are focused on older age and address risk-taking in several domains as indicated earlier. Risk-taking in life-critical systems requires high levels of skills, knowledge, expertise, maturity of practice, courage, creativity, familiarity, availability, adaptability, dependability, and boldness (Boy, 2014). This means a high level of work ability, which should be extensively studied and redefined for life-critical systems, requiring longitudinal studies in different contexts. Therefore, it is important to identify age-related attitudes and behavior patterns, particularly in relation to risk exposure and risk-taking across aging.

According to the Survey of Health, Aging and Retirement in Europe (Bonsang & Dohmen, 2015), the age-related change in attitude to risk is associated with cognitive aging. The results of this survey reveal that the process of cognitive decline is closely connected to changes in risk preferences over the course of life. Aging and the related declines could represent a leading factor in the decrease in performance and the perception of a risk above one's available resources. This is the result of a self-assessment based on the balance between the demands imposed by the situation at hand and the individual resources available to accept the risk and perform the corresponding actions. However, as indicated earlier, age-related declines do not have a negative impact on working performance during active life provided that the individual skills and maturity of practice, together with lifelong training and creativity, allow for self-regulation based on compensating for any declines. Therefore, high competencies, skills, knowledge, and maturity of practice, which are not affected by aging during an active life, are the basis for good judgments in risk assessment and decision-making. Additionally, risk-taking requires familiarity with the technology, the allocated tasks, the organizational constraints and possibilities, and the work environment, together with adaptability, availability, and dependability on the job and context. These job requirements, featuring life-critical systems, prepare operators to deal with the unexpected, which comes to be considered as a common situation rather than an exception (Boy, 2014).

9.6 FINAL REMARKS

Life-critical systems, being automated, require highly skilled, competent, and adaptable operators to deal with uncertainty, risk, and the unexpected. In an automated system, an early identification of a dysfunction requires the operator to switch from an automated mode to a manual mode, which requires higher competency levels mainly based on a heightened level of knowledge. Life-span education and training policies should ensure the continuous adaptability of older workers to technological development, making it possible to avoid resistant attitudes to new developments. However, the following aspects must be taken into account: (1) the importance of highlighting higher levels of knowledge toward a perfect understanding of how the system functions instead of basing training on skill acquisition; (2) the importance of basing learning and acquisition of new skills on previous experience and knowledge, which is the prime request for the success of learning and will eliminate any feeling of anxiety or reaction to change; and (3) the importance of a human-system integration approach in the system design to ensure the system's user-friendliness, and thus provide an easy appropriation of tools and efficient human-machine cooperation.

Life-critical systems have been widely studied, but there is a lack of information in literature regarding: (1) age-related performance in highly demanding work, and (2) the extent to which a healthy worker and safe behavior guard against safety-critical outcomes. Therefore, there is a need to:

- Establish minimum standards for work ability in specific contexts;
- Identify variables that mediate the relationship between age and performance, particularly in life-critical systems;

- Define criteria to assess age-related performance in particular jobs (life-critical systems);
- Identify the relevant factors to be incorporated into tools (cognitive and emotional demands, expertise, experience, task demands, transient factors, risk-taking attitudes, and behavior).

Although operators of life-critical systems will age during their active life, and their age-related decline could have an impact on task performance, there is a need to understand how aging affects behavior and task performance within life-critical systems. Risk-taking attitudes and behavior, as well as the influence of transient factors on decision-making, should be studied in close to real settings. Thus, the following research questions arise from the identification of research needs:

- Which age-related changes result in decreased performance at work?
- Which decreases in performance are safety-critical?
- Could some decreases in performance be safety-critical in a particular job?
- Which existing tools could be applied to assess human functioning and performance?
- What new tools should be developed to assess human functioning and performance?
- Should the abovementioned strategies to stimulate older operators' cognitive functioning be applied as a way of containing or recovering from cognitive declines?

The system safety and efficiency rely on people being able to cope with local uncertainty and variability factors while maintaining enough awareness of how their actions and decisions affect the system as a whole. Thus, the impact of aging in task performance under technical and organizational constraints should be investigated, particularly regarding: (1) the evolution of risk-taking attitudes and behavior across aging, (2) decision-making in the face of uncertainty and under time pressure, (3) the influence of expertise and maturity of practice in self-regulation of behavior and risk-taking, and (4) the effects of transient factors in risk-taking. Such research should allow the identification of the relevant variables for the development of tools and methods to assess and promote work ability among professionals within life-critical systems.

As a final statement, higher educational systems should evolve with a view to matching new societal needs that were already highlighted in the 1998 World Conference on Higher Education: "Towards an Agenda 21 for Higher Education". According to a publication from this conference, (UNESCO, 1998), "High education is facing the challenge of being prepared to adequately carry out its mission in a changing world and to respond to the needs and requirements of the XXI century society, which is revealing itself as the Society of Knowledge, Information Technology and Education". Some higher education and research institutions around the world have adopted new paradigms supporting updated programs and contents. In the USA, STEM (Science, Technology, Engineering, and Mathematics) Education (Boy, 2013; Leshner & Scherer, 2018) has become "a topic of discussion emphasizing that competitiveness is dependent on a strong educational program aiming at preparing innovative scientists and engineers who will provide innovations that will be vital for a flourishing economy in this technological era". While these initiatives are a wonderful start to the exploration of these four areas of study, the critical process of creativity and innovation is lacking. Thus, STEAM Education (Boy, 2013; Khine & Areepattamannil, 2019), appeared as an educational approach to learning through Science, Technology, Engineering, the Arts, and Mathematics as access points to guide student inquiry, dialogue, and critical thinking. Thus, the step from STEM to STEAM was short and fast, having taken root over the past several years as a positive strategy to meet the needs of a 21st century economy. STEM alone lacks several key components that are required today for safety-critical systems, particularly those in which life is at stake.

In Europe, the Finnish University AALTO was created in 2010 with the mission of shaping the future "building a competitive edge by combining knowledge from different disciplines to identify and solve complex challenges and to educate future visionaries and experts". Thus, science and art were put together with technology and business to accomplish the assigned mission. Now, time is running, and the future will not wait. It is there with new societal needs and an open labor market waiting for highly competent, prepared, and motivated innovators to shape our future.

NOTE

1 Retired Full Professor from Lisbon University – FMH.

REFERENCES

Bellet, T., Mayenobe, P., Bornard, J.C., Paris, J.C., Gruyer, D. et al. (2011). Human driver modelling and simulation into a virtual road environment. In Cacciabue, P.C.; Hjälmdahl, M.; Luedtke, A. & Riccioli, C. (Eds.), *Human Modelling in Assisted Transportation: Models, Tools and Risk Methods.* Springer, Milan, pp. 251–262.

Bolstad, C.A. (2001). Situation awareness: does it change with age? *Proceedings of the Human Factors and Ergonomics Society Annual Meeting 2001*, 45: 272.

Bonsang, E. & Dohmen, T. (2015). Risk attitude and cognitive aging. *Journal of Economic Behavior & Organization*, 112, pp. 112–126.

Borys, D. (2009). Exploring risk-awareness as a cultural approach to safety: exposing the gap between work as imagined and work as actually performed. *Safety Science Monitor*, 13(2), pp. 1–11.

Boy, G.A. (2003). Documenter un artefact. In Boy, G. (Ed.), *Ingénierie Cognitive.* Lavoisier, Paris.

Boy, G.A. (2013). *Orchestrating Human-Centered Design.* Springer, London.

Boy, G.A. (2014). Dealing with the unexpected. In Millot, P. (Ed.), *Risk Management in Life Critical Systems*, Wiley-ISTE, London, UK, November.

Boy, G.A. (2016). *Tangible Interactive Systems: Grasping the Real World with Computers.* Springer Nature, Cham, Switzerland.

Boy, G.A. (2020). *Human-Systems Integration: From Virtual to Tangible.* CRC Press, Boca Raton, FL.

Buljan, A. & Saphira, Z. (2005). Attention to production schedule and safety as determinants of risk-taking in NASA's decision to launch the Columbia Shuttle. In Starbuck, W.H. & Farjouin, M. (Eds.), *Organization at the Limit: Lessons from the Columbia Disaster.* Blackwell Publishing, Malde, MA.

Caserta, R.J. & Abrams, L. (2007). The relevance of situation awareness in older adults' cognitive functioning: a review. *European Review of Aging and Physical Activity* 4, pp. 3–13.

Chapon, A., Gabaude, C. & Fort, A. (2006). Défauts d'Attention et Conduite Automobile: Etat de l'Art et Nouvelles Orientations pour la Recherche. *Synthèse INRETS N. 52.* IFSTTAR, Lyon, France.

Czaja, S. & Moen, P. (2003). In Pew, R. W. & Van Hemel, S. B. (Eds.), *Technology for Adaptive Aging.* The National Academies Press, Washington, DC.

Desmond, P.A: and Hancock, P.A. (2001). In Hancock, P.A. and Desmond, P.A. (Eds.) Stress, Workload and Fatigue, pp. 455–465. LEA Publishers, London.

Eby, D., Shope, J.T., Molnar, L.J., Vivoda, J.M. & Fordyce, T.A. (2000). *Improvement of Older Driver Safety Through Self-evaluation: The Development of a Self-Evaluation Instrument.* UMTRI Technical Report 2000-04.

Eby, D., Molnar, L. & Kartje, P.S. (2009). *Maintaining Safe Mobility in an Aging Society.* CRC Press, T&F Group, Boca Raton, FL.

Endsley, M.R. (1995). Toward a theory of situation awareness in dynamic systems. *Human Factors*, 37(1), pp. 32–64.

Fu, L., Kessels, R.P.C. & Maes, J.H.R. (2020). The effect of cognitive training in older adults: be aware of CRUNCH. *Aging, Neuropsychology, and Cognition*, 27(6), pp. 949–962.

Holland, C.A. (2001). *Older Drivers: A Literature Review.* Report No. 25. Department for Transportation, London, UK.

Huang, Y., Wood, S., Berger, D. & Hanoch, Y. (2013). Risky choice in younger versus older adults: affective context matters. *Judgment and Decision Making*, 8(2), pp. 179–187.

Ilmarinen, J. (2005). *Towards a Longer Worklife: Ageing and the Quality of Worklife in the European Union.* Finish Institute of Occupational Health, Helsinki.

Khine, M.S. & Areepattamannil, S. (Eds.) (2019). *STEAM Education: Theory and Practice.* Springer, New York, NY.

Jackson, S. (2010). *Architecting Resilient Systems: Accident Avoidance and Survival and Recovery from Disruptions.* John Wiley & Sons, Hoboken, NJ.

Lee, J.D., Young, K. & Regan, M. (2009). Defining driver distraction. In Regan, M., Lee, J.D. & Young, K. (Eds.), *Driver Distraction: Theory, Effects and Mitigation*. CRC Press, New York, NY.

Leshner, A. & Scherer, L. (Eds.) (2018). *Graduate STEM Education for the 21st Century*. The National Academies Press, Washington, DC.

Fu, L., Kessels, R.P.C. & Maes, J.H.R. (2020) The effect of cognitive training in older adults: be aware of CRUNCH. *Aging, Neuropsychology, and Cognition*, 27(6), pp. 949–962, doi: 10.1080/13825585.2019.1708251

Mathews, G., Desmond, P., Neubauer, C. & Hancock, P. (2012). An overview of operator fatigue. In Mathews, G., Desmond, P., Neubauer, C. & Hancock, P. (Eds.), *The Handbook of Operator Fatigue*. Ashgate, Burlington, VT.

Mathews, G. & Desmond, P. (2001). Stress and driving performance: implications for design and training. In Hancock, P. & Desmond, P. (Eds.), *Stress, Workload and Fatigue*, LEA Publishers, New Jersey.

Oxley, J., Charlton, J., Fildes, B., Koppel, S., Scully, J., Congiu & M., Moore, K., (2005). *Crash Risk of Older Female Drivers*. Monash University Accident Research Centre, Clayton, Australia.

Rasmussen, J. (1987). Cognitive control and human error mechanisms. In J. Rasmussen, K. Duncan & J. Leplat, (Eds.), *New Technology and Human Error*. John Wiley & Sons, Ltd, Hoboken, NJ.

Reason, J. & Hobbs, A. (2003). *Managing Maintenance Error*, Ashgate, London, UK.

Rolison, J., Hanoch, Y., Wood, S. & Liu, P.J. (2013). Risk-taking differences across the adult life span: a question of age and domain. *Journals of Gerontology, Series B: Psychological Sciences and Social Sciences*, 69(6), pp. 870–880.

Simoes, A. (2003). The Cognitive Training Needs of Older Drivers. *Recherche Transport Sécurité*, no. 79, Avril-Juin.

Simoes, A., Pereira, M. & Panou, M. (2011). Transversal perspectives on human-machine interaction: the effect of age in human-machine systems. In Boy, G. (Ed.), *The Handbook of Human-Machine Interaction*. Ashgate, Oxfordshire, UK, pp. 79–94.

Van Zomeren, A.H. & Brouwer, W.H. (1994). *Clinical Neuropsychology of Attention*. Oxford University Press, New York, NY.

Vertzberger, Y.Y.I. (1998). *Risk-Taking and Decision-Making: Foreign Military Intervention Decisions*. Stanford University Press, Stanford, CA.

Tschiersch, I. & Schael, Th. (2003). Concepts of Human-Centred Systems. In Brandt, D. (Ed.), *Human-Centred System Design – First: People; Second: Organization; Third: Technology*. (Retrieved from: http://www.ima-zlw-ifu.rwth-aachen.de/fileadmin/user_upload/INSTITUTSCLUSTER/Alumni/human-centred_system_design.pdf)

UNESCO, (1998). World Conference on Higher Education: "Towards an Agenda 21 for Higher Education", Paris. Working Document. (Retrieved from: https://unesdoc.unesco.org/ark:/48223/pf0000113603)

United Nations (2004). *World Population to 2300. Department of Economic and Social Affairs. Population Division*. New York. (Retrieved from: http://www.un.org/esa/population/publications/longrange2/WorldPop2300final.pdf)

Chapter 10

Risk-seeking in healthcare

A life-critical necessity[1]

Robert L. Wears[†]
University of Florida Health Science Center Jacksonville
Gainesville, Florida

CONTENTS

10.1 INTRODUCTION

Shouldn't healthcare be safe? At first, risk-seeking behavior in healthcare seems a contradiction. We would like to believe that interventions to try to reduce our long-term burden of the consequences of illness, injury, and death to avoid risk to the greatest extent possible. However, as is shown elsewhere in this work this can only be true for certain types of activities, and unfortunately healthcare includes many activities for which the safest strategy is a risk-seeking one. For example, in a great deal of emergency medicine, intensive care medicine, and trauma surgery, a safety-first strategy may lead to disaster while a risk-seeking strategy, although briefly highly risky, may result in a better outcome.

This circumstance is analogous to discussions of financial risk (nothing ventured, nothing gained) and its roots are quite old. In the New Testament, the parable of the talents describes a master who is leaving his house to travel, and so entrusts varying amounts of his property to his three servants. When he returns, he asks for an accounting from the servants; to respond that they have put those resources to work and have achieved a substantial return. The third servant, however, did not want to face the risk of loss and so buried the money in the ground and returned the full amount (but no gain) to his master. For this excessive conservatism, he was punished.

As a general proposition, there are no risk-free options in healthcare. Every elective surgery such as joint replacement entails a trade-off between short-term risk (e.g., death in surgery) and long-term benefit. There is a clear risk-free option available – not to have the surgery – but many find the risk-benefit calculus favorable.

10.2 PROGRESS IN HEALTHCARE

A good deal of progress in healthcare involves the sorts of trade-offs. For example, the traditional method of reducing a shoulder dislocation dates to Hippocrates time and is at least vaguely reminiscent of torture on the rack. The operator pulls steadily on the dislocated arm while an assistant holds the patient's trunk against that pull. This traction-encounter-traction must be maintained for some time to overcome the muscle spasm secondary to pain that keeps the joint from moving; but after sufficient time with sufficient force, the shoulder joint can be put back into place.

Although the Hippocratic method is generally successful, it is clearly an undesirable experience for patients and doctors alike. In recent years, the development of rapid-acting anesthetics has radically changed the approach toward managing this and other painful procedures. These new drugs induce anesthesia very rapidly (seconds to minutes) and wear off very rapidly (minutes), thus allowing brief procedures to be performed under sedation rather than on an awake and sensate patient. However, procedural sedation such as this carries with it low but nonzero risks of apnea (stopping breathing) and hypotension (circulatory shock), which may ultimately lead to death or irreversible brain damage. The bottom line here is that advances in care that improve the general experience of most patients may inevitably be accompanied by increases in risk that are heavily borne by a few.

10.3 EXAMPLES OF RISK-SEEKING
IN EMERGENCY CARE

This section provides three examples of risk-seeking in emergency care. All three examples involve what are known as airway problems: one concerns teaching about difficult airway management; another concerns violating procedures and taking risks for safety; and a third describes an airway case under desperate circumstances.

Airway problems are a huge issue in emergency care. If the patient's airway is obstructed, it leads to death or brain damage in a matter of minutes. Therefore, assuring an adequate airway is the first principle in any emergency, because if there is no airway, nothing else matters. This is commonly reduced in medical education to the "ABC" principle: airway, breathing, and circulation. Physicians in training are constantly enjoined to address airway (A) first and not to move on to other issues until they are certain the airway is adequate. While in most cases, simple observation is sufficient to establish that the airway is adequate, if it is not, the airway must be established by inserting a tube into the trachea (windpipe). This procedure requires a great deal of skill (and so is frequently rehearsed). It is always done under time pressure and, in emergency settings, may be required on short notice (i.e., with little opportunity to prepare). Because of the difficulty of the procedure and the heterogeneity of the circumstances under which it is done (for example, bleeding, face or jaw injuries, and vomiting) a wide variety of techniques and tools have been developed to manage the situation. The ultimate management technique is a surgical airway – an emergency tracheostomy – created by making a small incision in the front of the patient's neck just below the Adam's apple, carrying it down to the windpipe, and inserting a breathing tube into a hole in the windpipe. Because of the circumstances under which it is performed, and because of the difficult anatomy of the neck, this procedure itself carries considerable risk.

10.3.1 Case 1: Teaching about the difficult airway

As noted earlier, there are many options for managing airway problems, some of which gain complete control of the airway while others gain only partial or temporizing control. The choice of procedure is commonly made by assessing the "order of urgency" for establishing an airway. Table 10.1 displays typical principles associating urgency with actions to be taken.

Table 10.1 Principles guiding airway procedures

Urgency	Description of need, actions	Safety strategy
Impending airway	Airway needed in 30–60 minutes, perhaps more. Risk assessment done in advance, abort if too risky	Ultrasafe: risk avoidance
Urgent airway	Airway needed in 15 minutes or less. Risk assessment focuses on planning contingency actions	Managed safety: risk accepting
Forced to act	Airway needed in 1 minute or less. Situation is uncontrolled and getting worse, heroic improvisation may be required	Resilient safety: risk-seeking

10.3.2 Case 2: Violating procedures

A young man was brought to the emergency department by the police, shackled in a "hogtied" position, face down and bent backwards on the stretcher (see Figure 10.1). He is screaming loudly, irrationally, and struggling violently. The police state that he had been driving erratically and crashed at low speed into a bridge abutment. When officers approached, he attacked the officer and was shocked by a Taser at least twice.

This setting is underspecified. The possibilities include drug congestion, head injury, psychosis, hypoxia, acute toxic delirium, and a host of other conditions. In addition, Tasers have been suspected of causing sudden cardiac arrest in similar circumstances, and in fact, a Tasered patient had suffered an unexpected cardiac arrest in this emergency department one month prior.

There is a standard procedure for gaining control in situations like this, known as the "rapid sedation protocol" which involves intravenous dosing only (to avoid oversedation from the depot effects of slowly absorbed intramuscular injections). The protocol specifies giving small doses of two different sedatives at five-minute intervals until adequate sedation is achieved. However, sometimes the rapid sedation protocol is not so rapid; if sedation requires four small doses, the procedure would take 20 minutes. In this case, because of the patient's extreme violence, the intravenous line was not likely to last that long. In addition, the patient's face down in unusual posture would prevent him from being turned onto his back to establish an airway or do cardiac massage if he should arrest.

The ED staff began the rapid sedation protocol, since the usual procedure was not likely to be successful. In addition, they felt the patient's physical position was itself a risk since it prevented establishing an airway should one be needed. Therefore, they decided to use a different approach

Figure 10.1 The "hogtied" position of this patient.

(Gilardi, Guglielmetti, Perry, Pravettoni, & Wears, 2009). They used the drugs normally used for procedural sedation to rapidly induce unconsciousness, unshackled the patient, turned him onto his back, re-restrained him, and then allowed him to wake up.

This approach was strictly against the regulations governing procedural sedation, which specified a stable low-risk patient, a known diagnosis, and very specific pre-procedural and post-procedural monitoring. None of these conditions held in this circumstance, and the staff in this case discussed the fact that if something went wrong (for example, if the patient arrested from the Taser) that it would look like their rule violation had caused the problem.

They attempted to mitigate the risk of this off-design base procedure by moving to a resuscitation bay to perform it, preparing airway equipment, and assigning one physician to be prepared to establish an airway, if necessary. It is also important to note that in this situation, the lead physician and lead nurse were highly experienced and fairly senior in the organization; in addition, they had a long history of working together and a high level of reciprocal trust.

This novel procedure was successful, and although it entailed a brief period of risk, no adverse events followed. Ultimately, it was discovered that the patient had no drugs on board, only a small amount of alcohol, and had suffered no head or other injury. He had had a long history of bipolar disorder, had gone off his dedication and was in a severe manic phase of his illness.

Interestingly, this procedure – which was novel at the time – has become part of the informal "bag of tricks" used in emergency care, illustrating the phenomenon of "migration" in complex life-critical systems (Amalberti, Vincent, Auroy, & de Saint Maurice, 2006).

10.3.3 Case 3: Desperate circumstances

A 37-year-old man was brought to the emergency department in severe respiratory distress, saying that he had drunk hydrofluoric acid and thought he was going to die. Hydrofluoric acid is used in industry. It is infamous in toxicology; for example, it can dissolve glass and rocks. The cause is very deep, initially painless burns, followed by the death of the damaged tissue.

This circumstance represents the "forced to act" row of Table 10.1. The severe respiratory distress shows that airway obstruction is already present. The circumstances suggest that it will rapidly get worse, that there will be only one chance to intervene, and so the first option must be successful. Airway intubations are usually assigned to interns under supervision, in order for them to gain experience but the lead physician decided this was not appropriate in these extraordinary circumstances. Fortunately, this was during the day shift, so there were several highly experienced physicians available. She assigned the best one to prepare for intubation under direct vision, assigned a second to prepare the fiber-optic bronchoscope as a backup, and a third one to prepare for a surgical airway. When all three physicians had their equipment open and operational, the first position was directed to proceed with intubation under direct vision. Intubation was successful on the first attempt, and the patient was transferred to the intensive care unit, where he unfortunately died of acute lung injury on hospital day three.

10.4 WHAT UNITES THESE CASES?

What is similar about these three cases beyond that they all deal with airway issues? At an abstract level, the cases are similar in that they represent patients who present at a certain level of risk, with conditions that will lead their level of risk to gradually increase over time to extraordinarily high levels, but in whom a brief high-risk intervention holds forth the possibility of moving them to a much lower level of risk in the future. Thus, in these cases, there is a trade-off in accepting a brief period of high risk in the present, in order to achieve a long-term low-level of risk in the future. In other words, successful action in these cases must be risk-seeking.

Discourses about risk in healthcare are quite mixed. Healthcare tends to love the heroic narrative, in which daring doctors overcome overwhelming odds. The heroic narrative also has a negative form, in which cavalier uncaring doctors expose patients to unacceptable risks. These narrative forms are determined by outcome and subject to hindsight and outcome biases (Henriksen & Kaplan, 2003).

Although the heroic narrative is traditional in healthcare, a new discourse has developed in the last 25 years: a scientific-bureaucratic, rational, and technocratic vision of risk management that entails a fundamental denial of risk manifested by calls for "getting to zero" (that is, no adverse events) (Clancy, 2010) and a fascination with the proceduralized and standardization found in ultrasafe industries. It is ironic that this pursuit of ultrasafe healthcare itself carries risks. In many senses, safe performance requires exposure to risk in order to learn how to recognize the limits of control, how to navigate away from those limits, and how to make "sacrifice" decisions that trade-off lower level for higher level goals.

There is some additional empirical evidence that such a risk-seeking strategy can be beneficial in healthcare. Direct observations of untoward events in cardiovascular surgery and their association with the quality of patient outcomes has shown that the surgeons with the best results had just as many untoward events during the course of the operation as those with the worst, but that they had better means of detection and a greater repertoire of responses available to steer the subsequent course away from disaster (de Leval, Carthey, Wright, Farewell, & Reason, 2000). Similarly, aggregate data from the National Surgery Quality Improvement Projects Study showed that hospitals with the lowest mortality had just as many complications as those with the highest, but they were better at recognizing adverse events sooner and had better means of response and recovery (Ghaferi, Birkmeyer, & Dimick, 2009). Thus, a "no risk" strategy of eliminating all adverse event precursors may ultimately degrade a system's ability to respond effectively to those risks that inevitably remain: a certain level of judicious risk-seeking seems necessary for safe and effective performance.

NOTE

1 This chapter was written by Robert (Bob) Wears in the beginning of the book project. Bob died a few months later. He was so involved within the topic of risk-taking in medicine that we decided to keep his chapter as an important archive nurturing the purpose of this book.

REFERENCES

Amalberti, R., Vincent, C., Auroy, Y., & de Saint Maurice, G. (2006). Violations and migrations in health care: a framework for understanding and management. *Qual Saf Health Care, 15*(Suppl 1), i66–i71. doi: 10.1136/qshc.2005.015982.

Clancy, C. M. (2010). Getting to zero: new resources aim to reduce health care-associated infections. *Am J Med Qual, 25*(4), 319–321. doi: 10.1177/1062860610370395.

de Leval, M. R., Carthey, J., Wright, D. J., Farewell, V. T., & Reason, J. T. (2000). Human factors and cardiac surgery: a multicenter study. *J Thorac Cardiovasc Surg, 119*(4 Pt 1), 661–672.

Ghaferi, A. A., Birkmeyer, J. D., & Dimick, J. B. (2009). Variation in hospital mortality associated with inpatient surgery. *N Engl J Med, 361*(14), 1368–1375. doi: 10.1056/NEJMsa0903048.

Gilardi, S., Guglielmetti, C., Perry, S. J., Pravettoni, G., & Wears, R. L. (2009). *Changing horses in mid-stream: sudden changes in plan in dynamic decision problems.* Proceedings of the 9th International Naturalistic Decision-Making Conference (pp. 211–212), Covent Garden, London, UK, 4–6 June 2007.

Henriksen, K., & Kaplan, H. (2003). Hindsight bias, outcome knowledge and adaptive learning. *Qual Saf Health Care, 12*(Suppl 2), ii46–ii50.

Chapter 11

Risk-seeking and the paradox of variability for safety in healthcare

Resonance[1] with R.L. Wears' text

Lucie Cuvelier
Université Paris
Saint-Denis, France

CONTENTS

11.1 RISK-SEEKING AND VARIABILITY IN HEALTHCARE: THE CASE OF ANESTHESIA

"Healthcare includes many activities for which the safest strategy is a risk-seeking one" (Wears, Chapter 10 of this book). This is the case with anesthesia. Anesthesia, as a branch of the medical discipline, is viewed as being a pioneer in the field of patient safety (Cooper & Gaba, 2002; Gaba, 2000). It has all the characteristics of an extremely safe system with, in particular, a low probability of occurrence of serious adverse events (Amalberti, Auroy, Berwick, & Barach, 2005). In France in 2003, the rate of mortality linked to this medical practice was 1/145,000; that is, 10 times lower than in 1980 (Lienhart, Auroy, Péquignot, Benhamou, & Jougla, 2004; SFAR, 2003). It is generally agreed that anesthesia has become safer thanks to major advances in pharmacology, improvements in monitoring techniques, and professional commitment to practice standards (Amalberti et al., 2005;

DOI: 10.1201/9781003221609-11

Cullen, Bates, Leape, & the Adverse Drug Event Prevention Study Group, 2000; Gaba, 2000; Kohn, Corrigan, & Donaldson, 1999).

What is ironic is that "this pursuit of ultrasafe healthcare itself carries risks" (Wears, Chapter 10 of this book). Because anesthesia is not a therapeutic practice as such, it is often considered to be "risk-seeking" in itself (Gaba, Maxwell, & DeAnda, 1987). Admittedly, it is possible to quantify a priori the anesthetic risks taken to minimize it (via, for example, the classification of the American Society of Anesthesiologists). Nevertheless, every anesthetic situation remains one that is changeable and potentially risky: even during an operation considered to be straightforward, the intervention may unexpectedly devolve into a critical state. The variability in surgical acts, associated with the dynamic character of the complex processes underway results in anesthetists constantly bracing themselves and preparing for unforeseen acute events, by developing notably anticipative and adaptive strategies (Xiao, 1994). Some of these adaptations could be contradictory to norms, rules, and standard practices (Cuvelier & Falzon, 2010). This is the paradox of risk prevention in ultrasafe systems: being ultra-rigid, they do not allow any adaptation outside the framework of the anticipated rules (Amalberti, 2001; Dekker, 2003, 2017).

Thus, the field of anesthesia, more than many others, raises questions about risk-seeking and the link between variability and safety. These questions invite us to embrace how this paradox can be expressed in designs. On the one hand, risk-seeking and variability in practice can be completely unjustified and could increase the probability of errors and accidents. It is thus necessary and inevitable from a practical perspective to anticipate and formalize standard practices. "Work-as-imagined" (WAI), which includes procedures, guidelines, instructions, codes, programs, among others, is the basis of how this standard work is designed (Hollnagel, 2017). On the other hand, risk-seeking and variability in practice can also be equated with flexibility and adaptability to circumstances and local conditions that are important sources of safety. This variability refers to "how something is actually done" in specific, real cases, with each implementation being singular and different. This is what Hollnagel (2017) calls "work-as-done" (WAD). Conceptualizing the articulation between WAD and WAI is a foundation of systems safety design (Cuvelier, 2016; Cuvelier & Woods, 2019).

In his text, Chapter 10, Wears exposes three cases of airway issues that illustrate the variability in real anesthetic situations and risk-seeking in healthcare ((1) Teaching, (2) Violating procedures and (3) Desperate circumstances). In these three cases, "there is a trade-off in accepting a brief period of high risk now, in order to achieve a long-term low-level of risk in the future". Our chapter extends Wears' statement, by discussing the fourth case of airway issues, in a common "potential situation"(Cuvelier, 2011). It focuses on the issue of understanding how, in practice, anesthetists cope with uncertainty and construct long-term safe practices. It reveals the implications of variability for safety (Section 11.2) and illustrates the link between variability and safety in the long run, through developing competences (Section 11.3). This raises questions about how the paradox of variability and safety can be expressed in design, in line with the purpose of this book. Finally, it opens pathways for how ergonomists and designers might address this paradox by dealing with the gap between WAD and WAI (Section 11.4).

11.2 THE PARADOX OF VARIABILITY IN HEALTHCARE: THE FOURTH CASE OF "AIRWAY ISSUES"

To investigate the paradox of variability and safety in anesthesia, we relate an ergonomics-rooted study that was conducted in pediatric anesthesia, based on the clinical analysis of a complex (but foreseeable) decision regarding the airways (Cuvelier, 2011, 2019b). The underlying principles of this analysis were those of work analysis in an activity-oriented approach (Daniellou & Rabardel, 2005; Filliettaz & Billett, 2015). Twenty anesthetists from two French hospitals were interviewed using a method combining individual "thinking aloud" interviews and case-based simulation. The simulation case was devised by an anesthesia expert using a genuine patient file. It was a relatively complex case, involving surgery on a child suffering from a relatively rare, but known, disease (syndactyly in a two-year-old child with Apert syndrome). The simulation included photographs and laboratory results.

Results of this study reveal that access to airways was recognized as difficult by all anesthetists interviewed. One major risk in this surgery is indeed to fail to give patients the respiratory assistance needed for survival. All 20 anesthetists mentioned the risks of difficult intubation and spontaneously gave details regarding the strategy for airway control they were considering for the child's surgery. But the results also reveal that, to cope with this potential airway issue, several strategies (laryngoscopy, fibroscopy, laryngeal mask) were discussed and analyzed by each anesthetist and that anesthetists frequently selected two possible solutions. Last, but not least, the results of this study reveal that anesthetists sometimes selected a solution (laryngoscopy) that they themselves and experts viewed as riskier (12/20). Why?

One hypothesis could be that the safer technique (fibroscopy) is more expensive to implement: the equipment takes longer to prepare (availability of the fibroscope, disinfection, etc.) and requires specific technical skills. This hypothesis, based on efficiency (vs safety), is consistent with models that analyze activity in terms of compromise (e.g., the efficiency – thoroughness trade-off) or sacrifice judgment (Flageul-Caroly, 2001; Hollnagel, 2009; Woods, 2006). These concepts indicate that the decisions that are taken at the various levels of organizations reflect the fact that workers' multiple goals cannot all be met. Among these goals, safety and performance seem to be in conflict, incompatible with each other: hence, people have to make "sacrifices" or trade-offs between these two contradictory goals.

This hypothesis was, however, soon eliminated from our study. First, the arguments regarding the constraints of use were not mentioned during the interviews. During interviews, performance and safety often appear to be conflated into a unique goal: to ensure the best possible level of health for the patient, anesthetists all assert that they try to meet all goals. Indeed in healthcare, the objectives of "performance" (improving patient health) and those of safety (avoiding "degrading" the health of patients) are not clearly distinct, let alone opposed (Mesman, 2009). Second, and more importantly, in all cases of "risk-seeking", the safer solution of fibroscopy was also selected as the second choice; it is viewed as a fallback solution, in case the laryngoscopy technique does not work. In other words, all the participating physicians, including those who contemplated "perhaps trying" the laryngoscope, asserted that, before intubating the patient, they would ensure that the equipment (in particular, the fibroscope) was available and ready and that, should they themselves not be fully competent, experts in the use of that piece of equipment (an ENT or fibroscopy specialist anesthesiologist) were also available. To pursue the understanding of this variability, we wonder who selects a solution considered as riskier? (part III.2.3).

Before answering the question, it is important to note that the spontaneous usual reaction is to deplore the variability and "risk-seeking" observed in the study: "A new discourse has developed in the last 25 years: a scientific-bureaucratic, rational, technocratic vision of risk management that entails a fundamental denial of risk manifested by calls of 'getting to zero' (that is, no adverse events) (Clancy, 2010) and a fascination with the proceduralization and standardization found in ultrasafe industries" (Wears, Chapter 10 of this book). Conversely, one may admit that compliance does not eliminate variability in possible solutions, even in an ultrasafe system such as anesthesia. In our case, since rescue options or escape routes are provided for in the event of airway difficulties, the "risk-seeking" strategies are all the safer and more valuable.

In addition, analysis of decision criteria show that this "allowable variability" may conversely be a safety factor (Cuvelier, Falzon, Granry, Moll, & Orliaguet, 2012; Cuvelier, Falzon, Granry, & Orliaguet, 2017): we show for example that anesthetists aim to plan compliant solutions based on the resources (especially the "intrinsic" resources as skills) of the operators actually or potentially involved in these situations. In other words, in order to ensure patient safety, the strategies chosen are not only intended to be compliant ones but also be "controllable situations" for the team, and therefore, they reflect the skills, the know-how, and the individual and collective preferences of operators. Thus, to choose and design solutions, anesthetists articulated procedural knowledge relating to clinical risks (using guidelines, rules, and norms) and local knowledge regarding their own resources and those of their colleagues. As these intrinsic resources are variable, the preparation of the same anesthetic task by different physicians in different places can lead to varying solutions. Other research carried out in risky dynamic situations (e.g., fighter aircraft piloting) shows that, as long as the variability is within limits defined by standard practice, it is actually a safety factor. Indeed, it is this variability that allows safe work to be implemented, allowing everyone to operate within their own area of expertise (Amalberti, 1996).

11.3 THE PARADOX OF VARIABILITY FOR LONG-TERM SAFETY

Questioning who selects a solution considered as riskier (laryngoscopy), we found that selecting (or not) the laryngoscopy technique as the first choice is related to the anesthetists' seniority in the profession. Whereas the more senior anesthetists mostly opt for intubation by fibroscopy, the younger ones are more tempted to use intubation by laryngoscopy and use intubation by fibroscopy as a fallback solution. In other words, the younger physicians are more likely to select laryngoscopy, which is viewed as a riskier solution.

In their speech, it is difficult to explicitly grasp the arguments and reasons underlying the move toward a riskier solution (laryngoscopy). Those anesthetists who selected this option talk about "attempting", "trying [their] luck", "going for it" namely "making at least one attempt" with the "technique [they are] good at", the one "[they use] most often". Behind these verbatim excerpts, the shadowy figures of the anesthetists appear, with their competencies, habits, and abilities to execute this basic technical procedure (i.e., intubation by laryngoscopy) in the profession. In other words, for those anesthetists who select intubation by laryngoscopy as the preferred option, the object of the activity is no longer solely the here-and-now of a particular situation (productive activity). The focus is also on the doers of the activity, their experience, present, and future competencies. Their implicit issue is to develop the practice of carrying out this intubation procedure in a variety of situations that are always specific and arduous. These analyses are congruent with other works (Amalberti, 1996). The younger anesthetists display a more constructive activity that could correspond to a phase of exploration of the world, one that addresses the challenge of a variety of possible situations.

These short-term initiatives for the benefit of improved long-term risk management are all the safer and more valuable as risks are taken in full awareness of those risks. For all the participant anesthetists, rescue options or escape routes are provided for in the event of failure of the attempt to ensure the patient's health. The idea is that the physicians can construct a safe place for themselves, containing skills and structures, so that they can develop their own competences while keeping in mind their patient's safety. This is known as the zone of proximal development (ZPD) in activity theory (Bruner, 1996; Vygotski, 1980).

To resume, our case illustrates that "variability" and risk-seeking are not only a factor of flexibility and adaptability to local circumstances but also a way of competences and skills development. Here also, in the long run, "there is a trade-off accepting a brief period of high risk now, in order to achieve a long-term low-level of risk in the future" (Wears, this book). Actions that could be viewed as short-term risk-seeking are in fact ways of learning by doing: the younger participants' risk-taking is driven by competence development including both the technical skills and the meta-knowledge that are essential in risk management. In other words, the productive activity, aimed at patient care, also appears to be a "constructive activity", aimed at competence development and more effective risk management in the long term for future difficult patients (Rabardel & Samurçay, 2001). In the end, risk-seeking may increase long-term safety because it engenders the qualities of adaptability and capacity for adjustment that are necessary to deal with complexity, constantly changing technologies, and the current "intractable" systems (Hollnagel, 2014; Woods & Dekker, 2000). It makes crucial features of "resilience", such as "graceful extensibility", and a capacity for initiative possible (Woods, 2019).

11.4 GOING BEYOND THE PARADOX OF VARIABILITY AND SAFETY IN DESIGN

Distinguishing between variability and risk-seeking on the one hand and formalized, standardized practice on the other requires us to recognize that there is a gap between how work is done (WAD) and how work is imagined or thought of (WAI). Emerging debates and increasing numbers of publications (Braithwaite, Wears, & Hollnagel, 2016; Hollnagel, 2012) calling attention to this gap between WAD and WAI prompt us to investigate the links between these recent international studies and older concepts developed in Francophone countries around "actual work" and "prescribed work" (Leplat & Hoc, 1983; Ombredane & Faverge, 1955; Wisner, 1991). Indeed, the recognition of a gap between WAI and WAD has been, for over 60 years, a central premise of activity-centered Francophone ergonomics (Daniellou, 1996; Filliettaz & Billett, 2015). However, although the WAD/WAI distinction is a long-standing one, the debates it has given rise to are not closed, and not all the conclusions for design have been drawn nor discussed enough internationally (Clot, 2006; Cuvelier & Woods, 2019; Duraffourg, 2003; Maline & Guérin, 2009).

According to an activity-centered perspective, three possible positions may be taken by ergonomists regarding the gap (Béguin, 2007; Maline &

Guérin, 2009). They outline three ways of viewing the ergonomists' line of work and their contribution to design depending upon whether designers attempt to eliminate the gap, manage it, or enhance it. These three positions clarify the ergonomic position in design and define the approaches ergonomists can adopt to account for future activities from their inception (Béguin, 2007). These differing perspectives provide the broad characteristics of three possible goals pursued by ergonomists and designers faced with the gap between WAI and WAD to build safety (Cuvelier & Woods, 2019):

1. Eliminating the WAD/WAI gap, viewed as harmful both to operators (for whom the cost of bridging the gap is high) and system performance.
2. Managing the WAD/WAI gap, viewed as necessary leeway that could however become problematic and generate risks if extended outside acceptable limits.
3. Enhancing the WAD/WAI gap, viewed as the genuine reflection of human work, the foundation of development that should not be thwarted if workplace health and performance are to be promoted.

11.4.1 The gap: an unwanted feature to be reduced

From the first perspective, the WAD/WAI gap is harmful and must be eliminated. Over time, the gap has become "irreducible" or "incompressible", and ergonomists have gradually abandoned the notion of finally eradicating it. Instead, they have fallen back on, for want of anything better, a more realistic objective: reducing the gap. In all cases, the issue at stake is to bring the actual situation as near as possible to the norm. From the perspective of workers' health, managing the gap is viewed as costly and dangerous. It is a source of tough working conditions, risk-taking, or even suffering. Ergonomic interventions in design are thus aimed at understanding the gap and "setting up processes designed to reduce this gap (as removing the gap, in terms of the definition of work activities, is deemed to be an inappropriate term, given the definition of activity)" (Petit, Dugué, & Coutarel, 2009, p. 64). In terms of system performance and safety, the gap signals noncompliance with the ideal aimed for. Many ergonomic interventions and studies then explicitly seek to remove and/or reduce this gap; among these are both past studies (Sebillotte, 1991) and more recent ones that are part of innovative trends. Hence, for example, a high-reliability organization is said to be one in which the gap between WAD and WAI has been removed (Hartley, 2011), or a healthy safety culture exists when 'work-as-done' overlaps 'work-as-planned' (U. S. Department of Energy, 2013). The WAD/WAI gap is also said to be a marker of resilience: a considerable gap indicates poor resilience. It is thus necessary to measure and close the gap (Dekker, 2006). This viewpoint is also found in research rooted in the Francophone ergonomics tradition. Hence, for instance, Nyssen and Berastegui (2016, p. 38) argue that the work analysis methods developed as part of the Francophone theories of activity seek to "reduce the distance between work as prescribed (what they call the task) and work as performed (what they call the activity)".

According to Hollnagel (2017), two approaches may be taken to implement the reduction: (1) forcing WAD to comply with WAI – as in the Lean Management system, for example. This avenue is like traditional safety approaches; or (2) modifying and designing WAI so that it corresponds to WAD. The latter avenue, which involves giving a normative value to reality, is frequently found in ergonomics: WAD becomes the foundation for the principles of a new injunction, likely to generate work situations that may be as rigid as the previous ones. This particular viewpoint is part of a more general aim to *design for use*, common in the discipline, that involves anticipating future activities and modeling people's behavior at the same time as designing rules and tools (Béguin, 2007). Since an activity-preexisting regulatory, organizational, and technical framework is necessarily going to entrench the possible dispositions, it is preferable to know and anticipate future actual activities so as to devise an appropriate framework.

11.4.2 The gap: a buffer zone to be controlled

The conceptual development from an approach seeking to eradicate the gap, to an approach with the goal of managing the gap is significant: it entails viewing the various practices and deviations from the rules not only negatively (as errors, violations, noncompliance, etc.) but also positively, by recognizing the need for and the benefits provided by the regulations. Hence, it is inappropriate to eliminate or reduce the variability: "The difference between WAI and WAD should not be looked at simply as a problem that ought to be eliminated if at all possible. The difference should instead be seen as a source of information about how work is actually done and as an opportunity to improve work" (Hollnagel, 2017, p. 12).

Nevertheless, variability, although needed, must be kept under control; a checking rule to manage the gap is needed to prevent variability from running riot (Falzon, Dicioccio, Mollo, & Nascimento, 2013) since "the gap between expected working and real working is considered as one of the most important sources of risk" (Fadier & De la Garza, 2006). Hence, it is a case of monitoring and damping variability, with a view to finding means of strengthening the variability that leads to the desired ends while stifling the variability that leads to undesirable events (Hollnagel, 2008; Hollnagel, Nemeth, & Dekker, 2008; Lundberg, Rollenhagen, & Hollnagel, 2009). In other words, the issue is to control the WAD/WAI gap, to "manage it" with "full awareness and deliberation" (Mollo & Nascimento, 2013, p. 218) and to "to find effective ways of managing that to keep the variability of WAD within acceptable limits" (Hollnagel, 2017, p. 13).

This vision can easily be likened to those focusing on the notions of boundaries or buffer zones.[2] The general idea is that there should be an anticipated space between WAI and the specific WAD. This buffer zone should not be too small, so as to leave space for the adaptations and arrangements used to deal with the unexpected. Nevertheless, it should be anticipated, monitored, and supervised beforehand; in other words, it should be included within the WAI. From the outset, it is necessary "to enable both judicious regulatory leeway and an appropriate supervisory framework through providing zones earmarked for debating the trade-offs" (Nascimento, Cuvelier, Mollo, Dicioccio, & Falzon, 2014). Béguin (2007, 2014) uses the notion of "plasticity" to refer to this design-based approach. This orientation involves systems that are sufficiently versatile, flexible, and malleable to be compatible with activities in real-life situations that enhance technical performance in terms of both productive efficiency and health.

The notion of leeway, or room for maneuver, developed in France is emblematic of this second approach (Daniellou, 2004). Observing that design choices open and close entire avenues for future opportunities, this author recommends planning, from the outset, room for maneuver that makes several acceptable operating procedures possible. The issue then is both to open "possible forms of activity" (as opposed to defining a single best way, or ideal operating mode) at the same time as "making impossible other, too risky, operating procedures" (Daniellou, 2004, p. 360). This relatively vague concept has aroused English-language researchers' curiosity (Caroly, Simonet, & Vézina, 2015; Stephens, Woods, Branlat, & Wears, 2011; Woods & Branlat, 2010).

Regardless of the various definitions, one central dimension of the notion of rooms for maneuver is situational: rooms for maneuver fall within the here-and-now. Hence, they are related to productive activities (Rabardel, 2005) and seek to identify the boundaries between WAD and WAI. Although the issue is to show the relations between them and to seek to reconcile them, WAD and WAI are viewed as two distinct, clearly differentiated entities, or even often two opposing and conflicting domains. Extending the WAI area is to increase formalisms and regulated safety, and this is necessarily achieved at the expense of stakeholders' autonomy and of their adaptation skills; in other words, this reduces the range of possibilities of WAD (Amalberti, 2007; Morel, 2007; Morel, Amalberti, & Chauvin, 2008; Pariès & Vignes, 2007). This dual vision contributes to a twofold, alternative approach, widespread in the field of safety, which has already been hotly debated (Duraffourg, 2003; Norros, 2004; Weick, 1998). From this perspective, research questions focus upon trade-offs, concessions, and the right balance to be struck between two incompatible answers: choosing between an operational mode, that is planned and anticipated in WAI, and an operational mode that is improvised in WAD. They also consist in designing and clearly positioning the boundary between the two domains. Hence, even though there is an attempt to reconcile WAD and WAI (Braithwaite et al., 2016), notably through the notion of rooms for maneuver, the two concepts remain distinct and antagonistic, because any operation outside of WAI (including any leeway) is still viewed as harmful.

11.4.3 The gap: a breeding ground for valuable resources

This third option entails viewing the gap as the genuine reflection of human work, something that needs to be analyzed for what it is worth and that adds value (Fanchini, 2003; Maline & Guérin, 2009). This starting point first assumes that the gap is not empty, and it is necessary to go beyond the previously outlined dichotomous approach that views WAI and WAD as two different and opposite entities.

From this third perspective, disturbances and the "risk-seeking" strategies devised to deal with them are not located in a non-nominal area, nor even in a controlled space permitting freedom and leeway; instead, they are part of the nominal, everyday operations of normal behavior. They indicate structural contradictions that induce activities and development (Clot & Santiago-Delefosse, 2004; Engeström, 2000). This approach is supported by numerous studies that go well beyond the domain of safety. On the one hand, various studies focusing on the execution of a task show that even in situations characterized as normal or routine, sometimes viewed as highly regimented, uncertainty and disturbances are pervasive (Marescaux, 2007; Perrenoud, 1999, p. 124; Teiger & Laville, 1972). On the other hand, research focusing upon improvisation indicates that the latter cannot be considered as the spontaneous creation of innovation, in contrast to the controlled respect given to a musical score written in advance and compliance with rigid rules. In real life, improvisation mechanisms are more nuanced and considerably more complex (Weick, 1998). In the domain of jazz, as in that of organizational management, improvisation is in fact based upon a mixture of written score and spontaneity (Chédotel, 2005; Lorino, 2005; Tatikonda & Rosenthal, 2000). Rules and formal structures, just like memories of the past, are essential elements to sustain adjustments and to enable ad hoc action. Without these rules, it is impossible to confront the unexpected (Dien, 1998; Weick, 1998). A musician's activities, just like those of engineers or project leaders, involve the use of various gradations of types of improvisation, from interpretation to composition through variations and embellishment.

Since variability and deviations from the norm are basic constituents of development, it is clear that it is no longer a matter of eliminating the gap between WAD and WAI, but neither is it a matter of dividing it nor inducing compatibility, as in the second orientation. This third orientation is based upon the notion of dialogism. The dialogical principle posits that "two or several different 'logics' are brought together into one unit in a complex manner (complementary, competing, and antagonistic) without losing their fundamental duality" (Morin, 1982, p. 176). It entails thinking that "contradictions can be reconciled only through a third level, that which goes beyond the two opposing poles. Something resembling a totality, opening up new opportunities for a new level of development of totality rather than merely balancing two opposing elements" (Engeström, 2011, p. 172).

A great number of studies in ergonomics and occupational psychology can be interpreted and analyzed in this dialogical framework of the relationships between what is given (in particular, prescribed work arising from WAI) and what is created within the activity (within WAD) (Béguin, 2010; Clot & Béguin, 2004).[3] These studies go beyond the dual vision that opposes, on the one hand, the regulations and norms (WAI) and on the other hand, real-life activities at least partly determined by these norms (WAD) to move toward a dialogue that "produces structuring effects upon both terms" (Béguin, 2010, p. 128). In this dialogical framework, the analysis then focuses upon this contradictory unit that enables WAD and WAI to engage in dialogue. It is no longer a matter of construing the two poles in terms of equilibrium, addition, connection, or "even less as a compromise or blend, but rather in terms of 'going beyond', which is particularly difficult to achieve" (Engeström, 2011, p. 172). Based on the work carried out for 50 years by Francophone ergonomists, it is possible to outline three resources that have been designed within the framework of cultural historical activity theories to define what actually grows and deserves to be nurtured within the WAD/WAI gap: competences, collective work, and instruments (Cuvelier & Woods, 2019). These three conceptual resources seek to go beyond the twofold WAD/WAI analysis and provide practical means of capturing the dialogical units developing within the gap.

11.5 CONCLUSION: CONCEPTUAL RESOURCES FOR DESIGNING SAFETY

Since "variability" and risk-seeking are not only a factor of flexibility and adaptability to local circumstances but also a way to develop competencies and skills, the aim of eliminating or reducing the gap may, considering the other two, appear to be somewhat simplistic and limited. Indeed, it corresponds to what we call "the traditional approach to safety" and it is usually applied during design processes (Cuvelier, 2019a; Hollnagel, 2014). These approaches nevertheless remain relevant as a general "ergonomics first aid kit" (De Montmollin, 1996). Reflecting upon human beings and modeling work ahead of design processes so as to anticipate and design systems that are better suited to people's actual future activities is a first important step forward. Even today, many engineering-based models view prevention solely from the perspective of pure technical rationality and invoke the human factor only a posteriori, in terms of acceptability and resistance to change (Pécaud, 2010).

The second orientation is not incompatible with the first one: it involves designing flexible systems and rules that incorporate enough leeway to allow adaptations while keeping control and restricting these adaptations to an acceptable level. The aim is also to help workers to develop their adaptive capacity and their ability to make trade-offs so they can benefit from this room to maneuver. Many current studies and design processes are in line with this safety orientation, using strategies such as opening areas for discussion and participatory methods (Ciccone, Cuvelier, & Decortis, 2018; Cuvelier, Benchekroun, & Morel, 2017; Rocha, Mollo, & Daniellou, 2015).

The third orientation is more removed from prevailing practices in the field of safety and may appear to be more difficult to set up. It introduces numerous and complex issues and entails a radical change in the way people see their world in order to accept it as it is, and to work with it rather than against it. Currently, (at least[4]) three research strands are developing this dialogical orientation in the trends originating in activity ergonomics (Filliettaz & Billett, 2015; Teiger & Lacomblez, 2013): activity clinics (Clot, 2004), vocational didactics (Pastré, 2011), and ergonomics (Béguin & Rabardel, 2000; Rabardel, 1995). They have given rise to three fundamental notions (collective work, competences, and instruments) that are currently being investigated in France. These three conceptual resources seek to go beyond the twofold WAD/WAI constitute the cornerstone for moving forward in a design process that preserves and takes advantage of the gap.

NOTES

1 "Resonance" is a way in which we are relating with and to the world, "when something or someone touches us, makes us vibrate body and soul, and the world and we come out transformed" (Rosa, 2019).
2 These concepts are numerous. Related to the notion of boundaries, one finds the following terms: "acceptable performance boundary" (Cook & Rasmussen, 2005; Rasmussen, 1997), "margins of acceptable performance" (Amalberti, 2001), "boundary conditions tolerated by use" (Fadier & De la Garza, 2006). Related to the notion of zone, one finds the "normal zone of illegal operations" (Amalberti, 2009), the "compensation zone" (Miller & Xiao, 2007), the "marginal zone" (Rasmussen, 1997), the "space of functional possibilities for action" (Vicente, 1999), the "flexible region" (Woods & Wreathall, 2008), the "sphere of possible situations" (Cuvelier & Falzon, 2010), the "space of possible future activities" (Daniellou, 2004), or even the "range of operating procedures" (Daniellou, 1985).
3 This framework has been envisioned in many ways: "establishing-established dialectic" (Béguin, 2012), dialectic between "antecedent norms" and their "partial, local, centered re-determining" (Schwartz, 2000), dialectic between the "local and singular" and the "general" (Daniellou, 2000, 2008; Schwartz, 2009), dialectic of adherence and *desadherence* for lack thereof (Durrive, 2015; Schwartz, 2009).
4 Of course, this list is not comprehensive.

REFERENCES

Amalberti, R. (1996). *La conduite des systèmes à risques*. Paris: PUF, Coll. Le travail humain.

Amalberti, R. (2001). The paradoxes of almost totally safe transportation systems. *Safety Science, 37*(2–3), 109–126.

Amalberti, R. (2007). Ultrasécurité, une épée de Damoclès pour les hautes technologies. *Dossiers de la recherche, 26*, 74–81.

Amalberti, R. (2009). Violations et migrations ordinaires dans les interactions avec les systèmes automatisés. *Journal Européen des Systèmes Automatisés, 43*(6), 647–660.

Amalberti, R., Auroy, Y., Berwick, D., & Barach, P. (2005). Five system barriers to achieving ultrasafe health care. *Annals of Internal Medicine, 142*, 756–764.

Béguin, P. (2007). Taking activity into account during the design process. *Activités, 4*(4), 115–121.

Béguin, P. (2010). De l'organisation à la prescription: plasticité, apprentissage et expérience. In Y. Clot & D. Lhuiliez (Eds.), *Agir en clinique du travail* (pp. 125–139). Ramonville, France: ERES.

Béguin, P. (2012). Conception et développement. Appropriation, dialogues et sens du développement. In Y. Clot (Ed.), *Vygotski maintenant* (pp. 175–191). Paris: La Dispute.

Béguin, P. (2014). The design of instruments as a dialogical process of mutual learning. In P. Falzon (Ed.), *Constructive Ergonomics* (pp. 143–156). Paris: CRC Press.

Béguin, P., & Rabardel, P. (2000). Designing for instrument-mediated activity. *Scandinavian Journal of Information Systems, 12*, 173–190.

Braithwaite, J., Wears, R., & Hollnagel, E. (2016). *Resilient Health Care, Volume 3: Reconciling Work-As-Imagined and Work-As-Done*. Boca Raton, FL: CRC Press.

Caroly, S., Simonet, P., & Vézina, N. (2015). Marge de manœuvre et pouvoir d'agir dans la prévention des TMS et des RPS. *Le Travail Humain, 78*(1), 1–8. doi: 10.3917/th.781.0009

Chédotel, F. (2005). L'improvisation organisationnelle. Concilier formalisation et flexibilité d'un projet. *Revue française de gestion, 1*(154), 123–140.

Ciccone, E., Cuvelier, L., & Decortis, F. (2018). From causality to narration: the search for meaning in accident analyzes. *Paper presented at the Congress of the International Ergonomics Association*.

Clancy, C. M. (2010). Getting to zero: new resources aim to reduce health care-associated infections. *American Journal of Medical Quality, 25*(4), 319–321. doi:10.1177/1062860610370395

Clot, Y. (2004). *La fonction psychologique du travail*. Paris: P.U.F.

Clot, Y. (2006). Alain Wisner: un héritage "dispute". *Travailler, 15*(1), 185–198. doi:10.3917/trav.015.0185

Clot, Y., & Béguin, P. (2004). Situated action in the development of activity. *Activités, 1*(2), [En ligne], mis en ligne le 01 octobre 2004, consulté le 2027 février 2018. URL: http://journals.openedition.org/activites/1242. doi:10.4000/activites.1242

Clot, Y., & Santiago-Delefosse, M. (2004). La perspective historico-culturelle en psychologie. *Présentation. Bulletin de psychologie, 469*(57-1), 3–4.

Cook, R., & Rasmussen, J. (2005). "Going solid": a model of system dynamics and consequences for patient safety. *Quality and Safety in Health Care, 14*(2), 130–134. doi:10.1136/qshc.2003.009530

Cooper, J. B., & Gaba, D. M. (2002). No myth: anesthesia is a model for addressing patient safety. *Anesthesiology, 97*(6), 1335–1337.

Cullen, D., Bates, D., Leape, L., & the Adverse Drug Event Prevention Study Group (2000). Prevention of adverse drug events: a decade of progress in patient safety. *Journal of Clinical Anesthesia, 12*(8), 600–614.

Cuvelier, L. (2011). *De la gestion des risques à la gestion des ressources de l'activité. Etude de la résilience en anesthésie pédiatrique.* Thèse de doctorat en ergonomie, Cnam, Paris.

Cuvelier, L. (2016). *Agir face aux risques, regard de l'ergonomie* (Vol. 2016-01). Toulouse, France: Fondation pour une culture de sécurité industrielle, Collection Les Regards – Gratuitement téléchargeable sur: http://www.foncsi.org/.

Cuvelier, L. (2019a). Da segurança dos pacientes à resiliência dos sistemas de saúde: Estado da arte. *Ciência & Saúde Coletiva, 24*(3), 817–826. doi:10.1590/1413-81232018243.05062017

Cuvelier, L. (2019b). Taking risks to improve safety? Workplace learning in anesthesia. *Journal of Workplace Learning, 31,* 537–550. https://doi.org/10.1108/JWL-12-2018-0153

Cuvelier, L., Benchekroun, H., & Morel, G. (2017). New vistas on causal-tree methods: from root cause analysis (RCA) to constructive cause analysis (CCA). *Cognition, Technology & Work,* 1–18. doi:10.1007/s10111-017-0404-8

Cuvelier, L., & Falzon, P. (2010). Coping with uncertainty. Resilient decisions in anaesthesia. In E. Hollnagel, J. Pariès, D. Woods, & J. Wreathall (Eds.), *Resilience Engineering in Practice: A Guidebook* (pp. 29–43). Ashgate: Studies in Resilience Engineering.

Cuvelier, L., Falzon, P., Granry, J. C., Moll, M. C., & Orliaguet, G. (2012). Planning safe anesthesia: the role of collective resources management. *The International Journal of Risk and Safety in Medicine, 24,* 125–136.

Cuvelier, L., Falzon, P., Granry, J. C., & Orliaguet, G. (2017). Développement des collectifs de travail et développement de la sécurité: une étude sur les décisions à risque en anesthésie. *Psychologie du Travail et des organisations, 23*(3), 255–272. https://doi.org/10.1016/j.pto.2017.05.004

Cuvelier, L., & Woods, D. (2019). Sécurité réglée et/ou sécurité gérée: quand l'ingénierie de la résilience réinterroge l'ergonomie de l'activité. *Le Travail Humain, 82*(1), 41–66.https://doi.org/10.3917/th.821.0041

Daniellou, F. (1985). La conduite de processus chimique, présence et pression du danger In C. Dejours, C. Veil, & A. Wisner (Eds.), *Psychopathologie du Travail* (pp. 35–41). Paris: Entreprise moderne d'édition.

Daniellou, F. (1996). *L'ergonomie en quête de ses principes, Débats épistémologiques.* Toulouse: Octarès.

Daniellou, F. (2000). Préface: invitations à travailler. In Y. Schwartz (Ed.), *Le paradigme ergologique, ou un métier de philosophe.* Toulouse: Octarès.

Daniellou, F. (2004). L'ergonomie dans la conduite de projets de systèmes de travail. In P. Falzon (Ed.), *Ergonomie* (pp. 359–373). Paris: PUF.

Daniellou, F. (2008). Développement des TMS: désordre dans les organisations et fictions managériales. 2ème congrès francophone sur les troubles musculosquelettiques: de la recherche à l'action, Montréal, 18–19 juin.

Daniellou, F., & Rabardel, P. (2005). Activity-oriented approaches to ergonomics: some traditions and communities. *Theoretical Issues in Ergonomics Science*, 6(5), 353–357.

De Montmollin, M. (1996). *L'ergonomie* (3ème ed.). Paris: La Découverte.

Dekker, S. (2003). Failure to adapt or adaptations that fail: contrasting models on procedures and safety. *Applied Ergonomics*, 34(3), 233–238.

Dekker, S. (2006). Resilience engineering: chronicling the emergence of confused consensus. In E. Hollnagel, D. D. Woods, & N. Leveson (Eds.), *Resilience engineering: Concepts and precepts* (pp. 77–92). Aldershot, UK: Ashgate.

Dekker, S. (2017). *The Safety Anarchist: Relying on Human Expertise and Innovation, Reducing Bureaucracy and Compliance*. Boca Raton, FL: Routledge.

Dien, Y. (1998). Safety and application of procedures, or how do "they" have to use operating procedures in nuclear power plants? *Safety Science*, 29(3), 179–187.

Duraffourg, J. (2003). S'engager à comprendre le travail. In C. Martin & D. Baradat (Eds.), *Des pratiques en réflexion* (pp. 513–532). Toulouse: Éditions Octarès.

Durrive, L. (2015). *L'expérience des normes – Comprendre l'activité humaine avec la démarche ergologique*. Toulouse: Octarès Éditions.

Engeström, Y. (2000). Activity theory as a framework for analyzing and redesigning work. *Ergonomics*, 43(7), 960–974. doi:10.1080/001401300409143

Engeström, Y. (2011). Théorie de l'Activité et Management. *Management & Avenir*, 42(2), 170–182.

Fadier, E., & De la Garza, C. (2006). Safety design: towards a new philosophy. *Safety Science*, 44(1), 55–73.

Falzon, P., Dicioccio, A., Mollo, V., & Nascimento, A. (2013). Qualité réglée, qualité gérée. In D. Huillier (Ed.), *Qualité du travail, qualité au travail*. Toulouse: Octarès.

Fanchini, H. (2003). Le métier d'ergonome: être ou ne pas être. In F. Hubault (Ed.), *Le métier d'ergonome*. Toulouse: Octarèse.

Filliettaz, L., & Billett, S. (2015). *Francophone perspectives of learning through work*. Basel, Switzerland: Springer International.

Flageul-Caroly, S. (2001). Régulations individuelles et collectives des situations critiques dans un secteur des services: le guichet de la Poste. Thèse de doctorat en ergonomie, EPHE-LEPC, Paris.

Gaba, D. M. (2000). Anaesthesiology as a model for patient safety in health care. *British Journal of Anaesthesia*, 320(7237), 785–788. doi:10.1136/bmj.320.7237.785

Hartley, R. S. (2011). High reliability organizations a practical approach. *CCRM Workshops and Seminars, Berkeley*.

Hollnagel, E. (2008). Resilience engineering in a nutshell. In E. Hollnagel, C. Nemeth, & S. Dekker (Eds.), *Resilience Engineering Perspectives: Remaining Sensitive to the Possibility of Failure* (Vol. 1). Aldershot, UK: Ashgate Studies in Resilience Engineering.

Hollnagel, E. (2009). *The ETTO Principle: Efficiency-Thoroughness Trade-Off: Why Things That Go Right Sometimes Go Wrong?* Farnham, UK: Ashgate.

Hollnagel, E. (2012). Resilience engineering and the systemic view of safety at work: why work-as-done is not the same as work-as-imagined. In *Bericht zum 58. Kongress der Gesellschaft für Arbeitswissenschaft vom 22 bis 24 Februar 2012, Dortmund* (pp. 19–24). Kassel, Germany: GfA-Press.

Hollnagel, E. (2014). *Safety-I and Safety-II*. London: CRC Press.

Hollnagel, E. (2017). Can we ever imagine how work is done? *HindSight - The ability or opportunity to understand and judge an event or experience after it has occured Eurocontrol, 25*, 10–13.

Hollnagel, E., Nemeth, C., & Dekker, S. (2008). *Resilience Engineering Perspectives: Remaining Sensitive to the Possibility of Failure* (Vol. 1). Aldershot, UK: Ashgate Studies in Resilience Engineering.

Kohn, L., Corrigan, J., & Donaldson, M. (1999). *To Err Is Human: Building a Safer Healthcare System*. Washington, DC: National Academy Press.

Leplat, J., & Hoc, J.-M. (1983). Tâche et activité dans l'analyse psychologique des situations. *Cahiers de psychologie cognitive, 3*(1), 49–63.

Lienhart, A., Auroy, Y., Péquignot, F., Benhamou, D., & Jougla, E. (2004). La sécurité anesthésique: où en est-on? Premiers résultats de l'enquête "mortalité" SFAR – INSERM. *Le Praticien en Anesthésie Réanimation, 8*(2-C1), 151–155.

Lorino, P. (2005). Un débat sur l'improvisation collective en jazz animé par André Villéger. In R. Teulier & P. Lorino (Eds.), *Entre connaissance et organisation: l'activité collective*. Colloque de Cerisy, Paris: La Découverte.

Lundberg, J., Rollenhagen, C., & Hollnagel, E. (2009). What-You-Look-For-Is-What-You-Find – The consequences of underlying accident models in eight accident investigation manuals. *Safety Science, 47*(10), 1297–1311.

Maline, J., & Guérin, F. (2009). L'ergonome: organisateur du travail ou travailleur de l'organisation? 44ème congrès de la Société d'Ergonomie de Langue Française (SELF), 22–24 septembre, Toulouse, France, 245–253.

Marescaux, P. (2007). Exigences, incertitude et ajustement des conduites. *Le Travail Humain, 70*, 251–270.

Mesman, J. (2009). Channeling erratic flows of action: life in the neonatal intensive care unit. In C. A. Owen, P. Béguin, & G. Wackers (Eds.), *Risky Work Environments: Reappraising Human Work within Fallible Systems* (pp. 105–128). Aldershot, UK: Ashgate.

Miller, A., & Xiao, Y. (2007). Multi-level strategies to achieve resilience for an organisation operating at capacity: a case study at a trauma centre. *Cognition, Technology & Work, 9*, 51–66.

Mollo, V., & Nascimento, A. (2013). Pratiques réflexives et développement des individus, des collectifs et des organisations. In P. Falzon (Ed.), *Ergonomie Constructive* (pp. 207–221). Paris: PUF.

Morel, G. (2007). Sécurité et résilience dans les activités peu sûres: exemple de la pêche maritime. *Thèse de doctorat en ergonomie*, Université de Bretagne Sud.

Morel, G., Amalberti, R., & Chauvin, C. (2008). Articulating the differences between safety and resilience: the decision-making process of professional sea-fishing skippers. *Human Factors: The Journal of the Human Factors and Ergonomics Society, 50*, 1–16.

Morin, E. (1982). *Science Avec Conscience*. Paris: Fayard.

Nascimento, A., Cuvelier, L., Mollo, V., Dicioccio, A., & Falzon, P. (2014). Constructing safety: from the normative to the adaptive view. In P. Falzon (Ed.), *Constructive Ergonomics* (p. 294). Boca Raton, FL: CRC Press.

Norros, L. (2004). *Acting under Uncertainty. The Core-Task Analysis in Ecological Study of Work*. Espoo, Finland: VTT Publications

Nyssen, A. S., & Berastegui, P. (2016). Is system resilience maintained at the expense of individual resilience? In J. Braithwaite, R. Wears, & E. Hollnagel (Eds.),

Resilient Health Care III: Reconciling Work-As-Imagined and Work-As-Done (pp. 37–47). Boca Raton, FL: CRC Press.

Ombredane, A., & Faverge, J.-M. (1955). *L'analyse du travail*. Paris: PUF.

Pariès, J., & Vignes, P. (2007). Sécurité, l'heure des choix. *La Recherche, 413* (Suppl. no 413), 22–27.

Pastré, P. (2011). *La didactique professionnelle: Approche anthropologique du développement chez les adultes*. Paris: PUF.

Pécaud, D. (2010). *Ingénieries et Sciences Humaines, la prévention des risques en dispute*. Paris: Lavoisier, collection Sciences du risque et du danger.

Perrenoud, P. (1999). Gestion de l'imprévu, analyse de l'action et construction de compétences. *Education Permanente, 140*(3), 123–144.

Petit, J., Dugué, B., & Coutarel, F. (2009). Approche des risques psychosociaux du point de vue de l'ergonomie. In L. Lerouge (Ed.), *Risques psychosociaux au travail* (pp. 51–72). Paris, France: L'Harmattan.

Rabardel, P. (1995). *Les hommes et les technologies. Approche cognitive des instruments contemporains*. Paris: Armand Colin.

Rabardel, P. (2005). Instrument, activité et développement du pouvoir d'agir In P. Lorino & R. Teulier (Eds.), *Entre connaissance et organisation: l'activité collective* (pp. 251–265). Paris, France: La Découverte "Recherches".

Rabardel, P., & Samurçay, R. (2001). From artifact to instrument-mediated learning. *Paper presented at the International Symposium on New Challenges to Research on Learning, Organized by the Center for Activity Theory and Developmental Work Research, Helsinki*, March 21–23, 2001.

Rasmussen, J. (1997). Risk management in a dynamic society: a modelling problem. *Safety Science, 27*(2–3), 183–213.

Rocha, R., Mollo, V., & Daniellou, F. (2015). Work debate spaces: a tool for developing a participatory safety management. *Applied Ergonomics, 46*, Part A, 107–114.http://dx.doi.org/10.1016/j.apergo.2014.07.012

Rosa, H. (2019). *Resonance: A Sociology of Our Relationship to the World*. Hoboken, NJ: John Wiley & Sons.

Schwartz, Y. (2000). Le paradigme ergologique, ou un métier de philosophe.

Schwartz, Y. (2009). Produire des savoirs entre adhérence et desadhérence. In P. Béguin & M. Cerf (Eds.), *Dynamique des savoirs, dynamique des changements* (pp. 15–28). Toulouse: Octarès.

Sebillotte, S. (1991). Décrire des tâches selon les objectifs des opérateurs. De l'interview à la formalisation. *Le Travail Humain, 54*(3), 193–223.

SFAR. (2003). Sécurité anesthésique: ou en est-on? 45ème Congrès de la Société Française d'Anesthésie Réanimation. www.sfar.org

Stephens, R., Woods, D., Branlat, M., & Wears, R. (2011). Colliding dilemmas: interactions of locally adaptive strategies in a hospital setting. In E. Hollnagel, E. Rigaud, & D. Besnard (Eds.), Proceedings of the 4th International Symposium on Resilience Engineering (pp. 256–262). June 8–10, Sophia Antipolis, France.

Tatikonda, M. V., & Rosenthal, S. R. (2000). Successful execution of product development projects: balancing firmness and flexibility in the innovation process. *Journal of Operations Management, 18*(4), 401–425.

Teiger, C., & Lacomblez, M. (2013). *(Se) Former pour transformer le travail. Dynamiques de constructions d'une analyse critique du travail*. Laval: Presses de l'Université Laval.

Teiger, C., & Laville, A. (1972). Nature et variations de l'activité mentale dans des tâches répétitives: essai d'évaluation de la charge de travail. *Le Travail Humain, 35,* 99–116.

U. S. Department of Energy. (2013). *DOE Handbook: Accident and Operational Safety Analysis* (Vol.1: Accident Analysis Techniques). Washington, DC: U.S. Department of Energy.

Vicente, K. J. (1999). *Cognitive Work Analysis: Toward Safe, Productive, and Healthy Computer-Based Work.* Boca Raton, FL: CRC Press.

Weick, K. (1998). Improvisation as a mindset for organizational analysis. *Organization Science, 9*(5), 543–555.

Wisner, A. (1991). La méthodologie en ergonomie: d'hier, à aujourd'hui. *Performances Humaines et Techniques, 50* (Exposé fait à Montréal au xviie Congrès de la Société d'ergonomie de langue française (1990), ré-édition de 1995 in Wisner Alain, Réflexions sur l'ergonomie, Toulouse, Octarès, pp. 111–128), 32–39.

Woods, D. (2006). Essential characteristics of resilience. In E. Hollnagel, D. D. Woods, & N. Leveson (Eds.), *Resilience Engineering: Concepts and Precepts* (pp. 21–33). Aldershot, UK: Ashgate.

Woods, D., & Branlat, M. (2010). Basic patterns in how adaptive systems fail. In E. Hollnagel, J. Pariès, D. Woods, & J. Wreathall (Eds.), *Resilience Engineering in Practice: A Guidebook* (pp. 127–143). Ashgate: Studies in Resilience Engineering.

Woods, D., & Wreathall, J. (2008). Stress-strain plots as a basis for assessing system resilience. In E. Hollnagel, C. Nemeth, & S. Dekker (Eds.), *Resilience Engineering Perspectives. Volume 1: Remaining Sensitive to the Possibility of Failure.* Aldershot, UK: Ashgate.

Xiao, Y. (1994). *Interacting with complex work environments: a field study and a planning model.* PhD dissertation, University of Toronto.

Chapter 12

Managing and preventing operational risk-taking at design time

Stories from a human-centered design project

Sébastien Boulnois
Round Feather, Inc.
San Diego, California

Edwige Quillerou
INRS, French Research and Safety Institute for the Prevention of
Occupational Accidents and Diseases
Vandoeuvre-lès-Nancy, France

CONTENTS

AGENDA

This chapter was written in the first person to facilitate smooth and introspective reading of the work, and rapid grasp of the choices made by the designer in interaction with various experts. It is a story told by a project manager and human-centered designer, Sébastien Boulnois (me), who closely worked with the support of Edwige Quillerou, an occupational psychologist who greatly contributed to the project and very much helped in writing this chapter. It will examine designer's activity, and how designers take risks when making design choices. The project presented here was designed from the outset with the aim of being as close as possible to the real activity of users of the technical system being designed. The methodology was therefore based on a Human-Centered Design (HCD) approach, which consists of co-developing a technical system seriously considering inputs from subject matter experts and various other specialists, including specialists in human factors and work analysis. I have chosen to write this chapter in such a way as to allow readers to immerse themselves in the designer's world. The aim was to better address the stakes for the designer to better appreciate the complexity of a co-construction project with user experts and other specialists in the field of psychosocial analysis, thus making them more aware of and more empowered to act on the risks leading to the final design choices.

12.1 INTRODUCTION: FROM SYSTEMS REQUIREMENTS TO HUMAN NEEDS, CONCERNS, AND DILEMMAS. WHERE DO OUR MOTIVATIONS LIE?

From an early age I was interested in **inter-human** and human-machine interactions, and how systems could be exploited to enhance **human experience and emotional stimulations** (e.g., to music and light effects). I trained as an engineer and was particularly captivated by the concept of human-machine systems introduced in the third year of my degree. Up to that point, I had heard only about systems and their requirements; the human side of the equation was new to me and caught my attention.

Engineering is a multifaceted training program. During the first two years, I learned mechanical, chemical, industrial, electrical, "you-name-it-all" engineering. In my third – and final – year, the concept of human-machine systems was introduced as part of one industrial engineering class. Up to that point, I had heard only about systems and their requirements; the human side of the equation was new to me and caught my attention.

At the end of my master's degree, I was required to complete a six-month internship. At that time, my interest in the human side of the human-machine system subject was growing. One of my professors offered us an opportunity that I could not refuse: a Trans-Atlantic project at the Florida Institute of Technology, in the Human-Centered Design (HCD) Institute. The chance to work on aerospace-related concepts on the Space Coast was something that I could not pass up.

The next year, after presenting my report on my internship and being awarded my master's degree, I was invited back to the Florida Institute of Technology to start a PhD in HCD. I was to focus on expert commercial pilots' perceptions of aviation weather and associated decision-making. This domain involves the study and design of life-critical systems because pilots rely on multiple resources to maintain optimal levels of safety, efficiency, and comfort during a flight. The interaction between and innovation related to other aspects of HCD also means that design must account for risk-taking and risk-management (through research, prototyping, user testing). This research will be presented in this chapter.

Both my internship and PhD experiences shaped and refined my vision of how to ensure an optimal level of safety, efficiency, and comfort for people, and how I implemented it. The first challenge is to determine people's needs relative to their intents and activities, and to translate these by talking to people and turning their insights into tangible prototypes. These prototypes then serve as a basis for improvement, as will be described here.

I am now employed as a user researcher at Round Feather, Inc. In this role, I use HCD and a powerful emotions-driven methodology to expand my vision. This approach is not only about making sure that people's needs are covered in the context of their intents and activities, but I also need to account for their feelings, by resolving conflicting concerns (i.e., dilemmas).

In Section 12.2, I introduce Edwige Quillerou, who contributed to my design project by complementing my research approach with her perspective and methods. In Section 12.3, I describe my research efforts deployed during my design project and identify key moments of risk-taking and risk-management. In Section 12.4, I analyze these moments and draw lessons from them. In Section 12.5, I identify the dimensions that were grounded in the key moments and lessons learned, and propose a risk-taking and -management framework, or model, from a HCD perspective. In Section 12.6, I reflect on the methodology and stories developed in this chapter and summarize how risk was approached throughout this design project.

Before I introduce Edwige, I must briefly explain my vision of HCD, the approach I will cover in this chapter, to help the reader understand its elements and nuances in relation to Edwige's Human Activity Development theory. In Sections 12.3 and 12.4, I provide more detail on how this approach was applied within the design project presented in this chapter.

To me, HCD emphasizes the human as central to the system to be designed, be it a product, a service, an organization, or a process. The aim is to understand the context that the human will be operating within – with their goals, tasks, and activities – and the associated challenges that will be encountered. This approach makes it possible to correctly articulate the problem or the question to be addressed in the design project. Prototyping and human-in-the-loop simulations (HITLS) can then be used to validate or explore designs aiming to resolve the problem. In practice, these phases are often blended over several design iterations, because I tend to take a pragmatic approach (i.e., using several methods to achieve my research goal). Indeed, HCD is a perpetually renewing process, as there is always more to understand about humans and the contexts in which they operate. The design solutions obtained should always account for this ongoing evolution.

12.2 COLLABORATION WITH AN EXPERIMENTAL OCCUPATIONAL PSYCHOLOGIST – COMMITMENT TO RISK-TAKING IN A HUMAN-CENTERED DESIGN PROJECT

In this section, I introduce my coworker, who helped me to understand and integrate the concept of Human Activity Development in my research. Edwige Quillerou joined me during my design project and contributed to this chapter.

Edwige is a psychologist specializing in work and occupational health and safety research. She is involved in qualitative and "clinical" research-intervention at workplace to analyze "real" work and help with its transformation for health and safety issues. These interventions focus on the means implemented to help transform the work to improve conditions for the humans involved. Her methodology involves co-analysis of operators' work to design and develop new and specific approaches to occupational activity in safe workspaces, while also accounting for mental, social, and physical health dimensions.

With her colleagues (clinical psychologists – positioned as close as possible to the human to understand how they act and why they act in certain ways) she is currently testing and refining an interventional framework to develop new ways of designing and creating innovative workplaces that are safer and more meaningful in terms of work quality and quality of working life, developed through teamwork. To identify methodologies and tools to help workers, requires use of theories and concepts. The most appropriate ones are identified by working with and talking to technical engineers and industrial designers.

Most technology has some human input; either it is for use by humans, or it is created by humans. When researchers, engineers, and designers aim to understand or create something about or for humans, I rely on a theoretical view of the human being and need to be aware of this fact. The reason I need to be aware of our own framework and viewpoint is that any theory can only touch one part of reality. It is thus impossible to understand the whole reality in all its complexity. Many theories about humans exist, and psychologists choose the ones that align best with their goals as scientists, their values and their practitioner and/or researcher paradigm (as defined by the deontological ethics for psychology). Psychologists, especially those involved in research, bear in mind that I have an epistemological and theoretical choice, enabling us to be scientifically consistent and practically understandable.

When Edwige started to work as a psychologist, and specialized in research investigating interventions in work organizations, she chose to study a subject that makes sense to her, to have the most impact: to aid workers and managers to create optimal workplaces, and do their work in the best conditions, for a better world. For this reason, her goal, as a researcher, was not only to study to but also understand, or identify, human mysteries, but above all, to improve workplaces, focusing on human occupational activity. This is another way to understand the human dynamics process.

Edwige and I shared our viewpoints and frameworks to develop the methodology applied in the Onboard Weather Situation Awareness System (OWSAS) project (the aim of which was to understand how pilots deal with weather during a flight; the methodology is expanded in the next section), to achieve a shared objective: to design a better workplace for pilots.

From a theoretical perspective, I explain why in our collaboration, I see the human through a theoretical human-activity perspective, as defined by Vygotsky (1978) and Leontiev (1981), and adapted for work activity by Engeström (2000, 2005) and Clot and colleagues (Clot, 2009; Bonnemain, Bonnefond, Fontes & Clot, 2019). This theory of occupational activity is centered on psychological functions that integrate both objective and subjective aspects. It interlinks with and raises questions about its interdependence with other aspects of human activity, namely material, social, and physiological aspects. All these aspects of life are included in the following dynamic human activity functions:

- definition;
- analysis;
- role in a design project.

To contribute to the HCD approach I advocate, I worked within Edwige's human-activity framework (Quillerou & Boulnois, 2020). This framework was applied to aid and support development of the activity, allowing the technical design to closely mirror the pilots' real work world. I believe that working together created an opportunity to contribute to the HCD approach. I do not consider that psychologists analyze in isolation, rather they consider the team and support development of the design work. Working with a multidisciplinary team of psychologists and engineers on technical systems is a way to develop a new human-activity approach, and thus design better workplaces for the future. This perspective is in line with one of the principal missions of work in occupational prevention.

In an environment involving perpetual change, in this sample design project, Edwige, the psychologist (researcher in occupational psychology), and myself, the HCD researcher, have tried to design for a better world using the same goal-driven approach, even though our frameworks, methodologies, and practical goals are distinct. I provided the means to describe human occupational activity through its social aspects (using occupational psychology tools) and its technical aspects (using design tools), by integrating experts in this framework. I developed all methodological aspects together to fully integrate pilots in the process involved in the technical design project.

Thus, I focus on the psychosocial and sociotechnical perspectives of risk throughout the entire design process, and by situating the risks.

12.3 HUMAN-CENTERED DESIGN OF A 3D WEATHER APPLICATION FOR THE AVIATION COMMUNITY

The proposal of a risk-taking management framework in this chapter is grounded in research efforts that were conducted in my "Human-Centered Design of a 3D-augmented strategic weather management system for the aviation community" design project. In this section, I provide information on the context of this research, the research question, how I used HCD to conduct my research, and some of the key points where risk-taking and -management were crucial during this journey. I will be using my dissertation (Boulnois, 2018) as a reference throughout the section. I will try to keep the rest of this section as citation-free as possible, to allow smooth narration.

Commercial airline pilots' main mission is to conduct safe, efficient, and comfortable flights. However, weather often gets in the way, and there is nothing that can be done, other than to deal with it. The North American National Transportation Safety Board (NTSB) reports that 23% of accidents in the US national airspace system involve weather. Analysis of a few commercial accidents suggests that human factors such as situation awareness, shared situation awareness, and decision-making associated with weather as a factor can lead to fatal incidents. In the research presented here, the focus was mainly on convective weather (i.e., weather evolving in the three spatial dimensions, associated with hazards such as wind shear, turbulence, and hail). I chose this type of weather for two reasons: (1) during their training, pilots receive little in terms of weather-related information and skills, and (2) the current limitations of operational ground-air technologies and services affecting pilots' weather-related situation awareness and decision-making capabilities. I decided to learn more about the former reason to be able to meaningfully act on the latter one.

In the United States, efforts have been made by the FAA and academic and industrial entities to improve the accessibility and visualization of, as well as interactions with weather information, both for ground and aerial systems. These efforts led to the proposal of several visualization techniques (e.g., 2D and 3D). After looking at both the basic research and aviation-related research relating to these techniques, I realized that the content displayed, the dimensionality and the viewpoint used, as well as the interactive features available during the task to be performed would inform me on the optimal weather visualization techniques.

At that point, the research question I wanted to tackle was clear: how can weather information, presented in 2D and/or 3D, and the associated interaction features, affect airline pilots' weather-related situation awareness and decision-making capabilities?

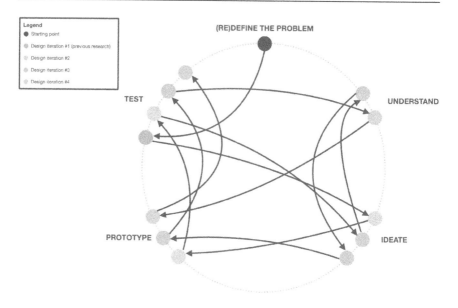

Figure 12.1 Design phases involved in the research.

I used HCD to navigate this research journey and answer the research question. This involved a lot of feedback obtained through interviews with expert pilots, ideation sessions with various team members, development of prototypes of weather applications, user testing, and other activities. The following diagram summarizes the phases involved (Figure 12.1).

First, I defined the research question. My next step was to study the previous internal weather-related research and systems that had been developed at my lab. The weather application resulting from these studies was named "Onboard Weather Situation Awareness System" (OWSAS); the term used from here onward. I reused these research elements and adapted them to make them the starting point to my journey. I then teamed up with a work analyst and two HCD interns to design an interview guide and immersed myself in the work of four expert pilots to gain an understanding of how they deal with weather during a flight. Based on these interviews, the interns and I developed weather representation solutions that we assessed in an ideation session conducted with two expert pilots. We then selected the optimal weather representation and developed a new OWSAS prototype, which was tested with five expert pilots. I analyzed the feedback and recorded the insights them for the next prototype iteration. During this user testing, the pilots presented emerging weather-related points.

Since my knowledge of the domain was growing as time went on, I decided to integrate these insights as questions in the immersion guide. Next, I used the guide to conduct interviews with 11 additional expert pilots. The purpose of these interviews was not only to learn about how pilots deal with weather during a flight, but also to determine how they interact with the various other stakeholders (i.e., air traffic controllers, dispatchers) to obtain the information they need to make decisions. Based on the first set of interviews, the insights from the user testing, and the second set of interviews, I refined my prototype and conducted a second round of user testing with seven expert pilots. Finally, I gathered the insights and used them to conclude my research.

Throughout my research journey, I was exposed to risk from several perspectives. In the next section, I describe the core context and some of the key moments and explain the lessons we learned.

12.4 RISK-TAKING AND -MANAGEMENT DURING MY DESIGN JOURNEY

12.4.1 Key moment 1 – interviewing pilots in flight conditions

As mentioned in the previous section, one of my main goals was to understand how pilots deal with weather during a flight. To do so, three paths could have been taken: (1) observe; (2) ask questions; (3) do both. Observing gives you the opportunity to analyze behavior in situ. Asking questions gives you the opportunity to understand why specific behavior is engaged. It is not uncommon to show operators video footage of themselves while asking contextualized questions, to reduce potential interpretations and bias. Therefore, adopting both methods would have been ideal to unpick what pilots do, how they do it, and why they do it. Unfortunately, it was impossible to observe pilots as they dealt with weather during a real flight, because it would have impacted the safety of the flight. Therefore, we chose option 2: ask questions. What follows is an explanation of experimentation with two interview configurations that involved risk-taking and -management.

I worked with Edwige and two HCD interns to design an interview guide for semi-structured interviews (i.e., prepared questions to impose a sequence on the conversation, while allowing some room for unexpected questions). We could not observe pilots' behavior during a real flight, but we believed that only asking questions in a nonflying environment would reduce the quality of the answers obtained. We hypothesized that interviewing pilots in flight simulators would bridge the gap between a real flight and a nonflying environment. At that time, I happened to be in charge of managing two high fidelity flight simulators. We therefore decided to conduct the interviews in the flight simulator (Figure 12.2). Consequently, the interview guide was constructed based on various phases of a flight (before a flight, at the airport, in the airplane, after takeoff). The idea was that the pilots would be asked relevant questions based on the flight phases as they dealt with them in the simulator.

Figure 12.2 During the first interviews with pilots (on the right, an expert pilot; on the left, the two student HCD designers).

Here is a brief overview of the protocol used: (1) expert pilots with experience flying the same type of airplane (i.e., B737) were recruited; (2) pilots were welcomed into the room containing the flight simulator; (3) the research and participants were briefly introduced; (4) the pilots entered the simulator; (5) the interview started. We gave pilots the green light to takeoff based on the sequence of our questions. The next part of the explanation is based on interviews conducted with the first two pilots (pilot #1 and pilot #2).

During the first and second phases (i.e., before the flight, at the airport), the interview with pilot #1 went smoothly. As we started asking questions relating to when pilots enter the cockpit of the airplane, he started to use the controls and visuals of the simulator to support his answers. This behavior increased when he took off and as we entered the after-takeoff phase. Pilot #1 started taking control of the simulator, making sure that the airplane would not crash, scanning the simulator instruments frequently to ensure that the variables were looking good. The pilot repeatedly said, *"Excuse me"*, *"Give me one second"*, *"Can you repeat the question?"* so that he had time to deal with the flight. These expressions were sometimes replaced by long silences. The interview took longer than expected. Pilot #1 became increasingly involved in scanning the instruments and ensuring the flight was safe, as we kept going through the questions. He finally attempted to land but crashed. Regardless of the reason, he appeared overwhelmed. We decided not to restart the simulation so that we could finish asking the remaining unanswered questions.

This experiment taught us a lot. First, flying is life-critical and requires the pilot to understand and use complex systems. The description presented by pilot #1 of the elements required to ensure a safe flight at the beginning of the interview supports this observation. Flying is also cognitively challenging and requires concentration. During the interview, we used the pilot's silences as an opportunity to ask follow-up questions to understand what was happening. Pilot #1 was scanning instruments every few seconds to maintain optimal situation awareness (Endsley, 1988), which is the key to ensuring a safe flight. These elements, combined with the final crash of the airplane in the simulation suggested that conducting an interview while a pilot performs critical work might not be the best approach: the session took almost twice as long as expected, and the pilot ended up being over-whelmed. With pilot #2, we maintained the same protocol but decided to pause the simulation as we entered each flight phase to reduce the cognitive load. Thus, the pilot did not have to deal with the flight and could focus more on the questions, improving the quality of the interview overall.

12.4.2 Key moment 2 – comparing two pilots' experience and knowledge

After we had interviewed four pilots in flight conditions, we understood that convective weather was their main concern. We then spent some time designing 3D visual representations for this type of weather information. We came up with four potential solutions and wished to pre-validate them before testing some in the flight simulator (Figure 12.3).

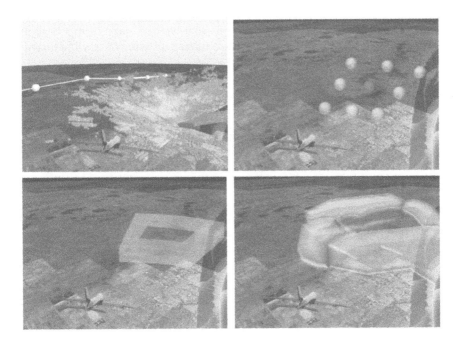

Figure 12.3 Weather representation solutions shown during an interview with two expert pilots.

Edwige Quillerou had introduced me to the idea of comparing pilots' opinions to obtain a better quality of insights, and this tied in with my initial idea of conducting work sessions with expert pilots to gather information on their perceptions and initial thoughts. Indeed, experts in general have knowledge and experience that they rely on to explain things. When several experts are involved in a single conversation, if the claims of one appear questionable to the others, it should trigger debates in real-time, leading to insights emerging that would not have been obtained if only one viewpoint was given to the research team. The idea sounded very interesting and exciting to me. I decided to give it a try and see what came of it. Together with Edwige Quillerou, we treated in detail the relevant key moments (Boulnois & Quillerou, 2020).

Several people took part in this working session: Edwige, one HCD intern, two expert pilots (pilot #1 and pilot #4), and me. We gathered everybody together in a room. The context and purpose of the session were explained to the pilots. The purpose from a research perspective was to compare our perceptions of the representations presented to what the pilots understood from them. We projected the four weather representation solutions so that the pilots could comment on them. Pilots #1 and #4 were given the floor.

The pilots started asking a lot of questions because they were confused about what they were looking at. After a period of reflection, they understood the representation related to convective weather. However, they still had many questions regarding the shapes used, the spaces between the shapes and other representation decisions that we had made from a design perspective. As shown on Figure 12.4, we started drawing on a white board to support my explanation.

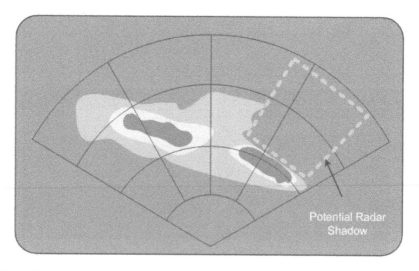

Figure 12.4 Attenuation phenomenon on a weather radar.

Figure 12.5 Designer using visual methods to convey his point to expert pilots.

The elements I drew (a rough representation of Florida, an airplane with some weather cells behind and in front of it) caught pilot #1's attention: "*I don't care what's behind me. [...] I might want to go in front and see what I can't see because of attenuation, but otherwise I want the representation where I am at and where I'm going*". In aviation, a common concern for pilots is the attenuation phenomenon (or radar shadow). Picture heavy weather in front of your airplane. Your radar sends out pulses in front of the airplane to sense the intensity of the weather, but none of the pulses can penetrate the first layer of the weather, and consequently they are reflected at your radar. You end up visualizing large red (sometimes magenta) cells on your screen, with nothing behind them (Figure 12.5).

I acknowledged his comment and told him that weather information would be available for the whole flight path because it was the purpose of the concept being designed. I also added that the attenuation phenomenon would no longer be an issue, for the same reasons, and that he would need to rotate around the first layer of weather (i.e., the red/magenta cells) to see what was behind it because it is a 3D view.

Pilot #1 was really confused. In contrast, pilot #4 remained silent, compiling all the information and explanations that had been tossed about in the room, during this time pilot #1 and I engaged in a rally to try to resolve our misalignment. After having presented multiple pictures, changes in words, placing several objects in a certain manner on the table to convey the point, we had not advanced that much – but then pilot #4 said "*Oh, I get it!*". He first played it back to me to make sure he had understood; then he used aviation-specific terms to convey the point to pilot #1. Pilot #1 clicked; we were finally all aligned on the concept and what we, as designers, were trying to achieve to help them, as pilots.

This situation taught us a few things. First, collaborative work involving experts from different fields requires terminology alignment. Designers have a distinct vocabulary to pilots. This can sometime lead to misunderstandings between the parties, as in the situation described. In this case, the use of a range of expression tactics (e.g., verbal, verbal + gestures, verbal + gesture + whiteboard/objects) seemed appropriate as they allowed the point to be conveyed from different angles, eventually leading to alignment for all participants. Second, having two pilots instead of one contributed to conveying the point. In this case, pilot #4 served as a bridge between pilot #1 and me, pilot #4 thus controlled and moderated the meeting for a period to achieve common understanding.

12.4.3 Key moment 3 – final prototype evaluation: modifying pilots' current radar interaction model to allow comparison with OWSAS

Overall, over a few months, we interviewed 15 expert pilots, built one functional prototype of OWSAS, and performed user testing. Results were obtained and analyzed. A final prototype was then built based on these results, ready to be evaluated. This key moment relates to the final user testing. For the sake of clarity, I will explain a few technical aspects of this knowledge management (KM).

The purpose of user testing was to compare current weather radars (2D representation of weather information) and OWSAS based on two main criteria: situation awareness and decision-making. It was thus important to have the same weather information displayed on both systems. In other words, if I were to display some weather on OWSAS, it would have to be identical on the weather radar installed in our flight simulator. Although OWSAS was functional from an interaction perspective (i.e., all iPad features and buttons were working), it could not synchronize with weather information from ground radars or a specific weather database. In contrast, the flight simulator was equipped with real-time weather capabilities (i.e., the real weather and flight simulator weather were synchronized). It was also equipped with weather presets (i.e., thunderstorm, fog, clouds, etc.). However, there was no way to retrieve this data in a format that could be read by OWSAS, nor to inject specific weather information into it. This aspect of the work was hampered by the fact that I was running out of time to adapt OWSAS to achieve these aims, due to restricted research resources and time constraints.

The workaround I found to this issue was to mimic the flight simulator's weather radar on an iPad, the same platform OWSAS was prototyped on. Figure 12.6 shows the configuration of the weather radar in the flight simulator and how it was translated for display on the iPad. Several key features had to be included in this replica: (1) the screen on which the weather is visualized (blue); (2) the range of potential weather, up to a certain distance, that can be visualized (green); and (3) the tilt that allows you to angle the radar antenna up or down to see what is above or below the current view, and the gain that allows the sensitivity of the radar antenna to be fine-tuned (orange).

While creating this work-around, I became aware of several things. First, potential tangibility issues could arise when pilots interacted with the replica. Indeed, mimicking the weather radar controls (orange and green

Figure 12.6 Flight simulator radar (left); iPad-based replica (right).

boxes) on an iPad turned rotating actions (i.e., turning the knob to adjust range, tilt and gain) into single-push and sliding actions. Second, spatial cue issues; grouping all interaction controls in one place next to the screen compared to their initial locations in the cockpit was a big change. I therefore expected the pilots to take some time to acquaint themselves with these changes in placement and interaction.

Six expert pilots were included in the experiment. Pilots were asked to rate their perceived confidence and experience with tablets before we started the experiment; this perception served as a criterion to account for during analysis of the results. I also briefed them about changes in interaction and positioning before the session so that they would still focus as much as possible on the main tasks in the scenario. The scenario was as follows: the airplane is on the ground at Miami international airport, and the pilots are waiting to takeoff; there is convective weather activity in the vicinity of the airport. The two main tasks were (1) to assess the weather situation, and (2) to make an informed decision about whether they should takeoff.

During the testing session, pilots did not feel uncomfortable using the replica, although they acknowledged the changes. Specifically, a few pilots mentioned that they would wait a few minutes to see how the weather evolves on the radar to gauge the dynamics. In the case tested, it never changed because my weather scenario was a static image. Two out of five pilots actually found the replica better than the current radar configuration provided in the simulator, which is a nice emerging insight. However, a challenge that I had not anticipated emerged regarding the decision-making process. Pilots relied heavily on what they could see outside the airplane in the simulator to make the decision to at least delay takeoff, rather than relying on the weather radars, whether provided by the replica or by OWSAS. They then told me that they were so used to flying from Miami international airport that they did not require any additional tools to make the right decision in this specific weather scenario.

This situation made me realize that I had worked with expert pilots throughout my research to understand how they deal with weather during a flight and the issues with aviation weather, to identify optimal weather representation solutions for integration into OWSAS and to evaluate OWSAS either alone or in comparison with other weather radars. However, I did not call on an expert pilot when designing the scenario. Consultation would have helped ensure that the scenario would be challenging even for expert pilots with thousands of flight hours. This error prevented me from understanding the impact that OWSAS would have on the pilots' situation awareness.

12.4.4 Key moment 4 – current research considers pilots only; the reality is different

When I established my research question, I set up specific boundaries for the field I was about to enter, even though I knew that there was more to consider. As I progressed toward my goal, following interviews with 15 expert pilots, I realized that the national airspace system is not only about them and the systems contained within the airplane they are flying, but also about the other agents supporting them. Figure 12.7 shows the AUTOS (Artifact, User, Task, Organization, Situation) pyramid that I used to capture the complexity of the environment I was learning about. The nodes cover five elements: the artifact(s) (e.g., OWSAS), the main agent(s) (e.g., pilots), the task(s) to be performed (e.g., fly, navigate, communicate), the organizational agents (additional agents that interact directly or indirectly with the main agents and system(s) – e.g., air traffic controllers, dispatchers), and the situation(s) (e.g., adverse weather). The edges describe the relations linking the nodes. They include ergonomics and training procedures (artifact-user), task/activity analysis (user-task), role/jobs analysis

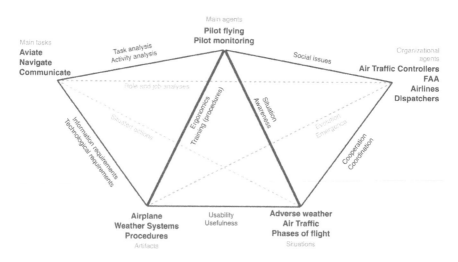

Figure 12.7 AUTOS pyramid applied to the OWSAS project. (From Boy, 2017.)

(task-organizational environment), cooperation and coordination (organizational environment-situation), and usability and usefulness (situation-artifact). I will briefly focus on a few organizational agents, as a full explanation of the whole picture would be complex.

Peers, dispatchers, and air traffic controllers can all play a significant role in helping pilots deal with weather during a flight. Before they get to the airplane, pilots are pre-briefed at the airport. They may also talk to other pilots coming back from a flight, during these conversations they sometimes discuss the weather around the airport. In the cockpit and/or in the air, pilots can request information from dispatchers relating to the weather in an area of interest, and from air traffic controllers regarding weather around their control towers. These agents have some systems that they can use directly or indirectly to support pilots' weather situation awareness. In my research, and especially during the design of my research plan, I decided to only interview, collaborate with, and test OWSAS on pilots because they would be the main agents interacting with the system. This decision was also made based on the time and resource constraints for my project (i.e., lack of network of air traffic controllers and dispatchers, limited response rate with respect to the research plan). In other words, although I was aware of the importance of the overall context, my focus was limited and only allowed me to investigate the pilots' point of view. We were nevertheless already considering the potential of OWSAS for dispatchers and air traffic controllers; how would it change the way they support pilots requesting help during a flight? How would it change their own activities (i.e., providing weather information and handling the aircraft's path)? When designing my experiments, I only considered pilots' interaction with OWSAS. What if dispatchers and air traffic controllers were also part of the picture? What if I included these agents as direct points of communication for the pilots? What if I gave these agents OWSAS too? How would they have behaved? Many questions remain unanswered.

Generally, any study requires the researcher to focus on a specific area of a field even though the whole picture needs to be accounted for. Thus, we are faced with a dichotomy: we want to delve deep in one area, but study of the whole picture requires a higher-level approach. In my case, I started with the study of pilots. In future research with OWSAS, I suggest that organizational agents and other elements highlighted by the AUTOS pyramid should be considered, as it will generate insights of better quality, raise more research questions, and speed the maturation of OWSAS.

12.4.5 Key moment 5 – performing early-stage research for life-critical systems

I would like to end this series of key moments with a moment that has no fixed point in time. Rather, this is something that I came to realize as I progressed with my research. As mentioned at the beginning of this chapter,

Figure 12.8 Overview of OWSAS.

and based on the literature I reviewed, a context-specific combination of weather information visualized in 2D vs. 3D associated with iPad-based interaction features was new. Figure 12.8 shows an overview of OWSAS. I will start with the content and will then explain how it was presented, before finishing with the interaction controls.

As far as content is concerned, OWSAS included a map, information on weather, altitude, flight path, and the airplane itself. The map was retrieved from NASA servers. The weather information was simulated using a web-based drawing method. Altitude level information was also simulated based on the simulated weather. The flight path information was generated by the flight simulator and was transferred to OWSAS. The airplane information was synchronized based on the flight simulator's location.

All the information was presented in 2D via a bird's-eye view, and in 3D via an exocentric view. The weather information was represented by green, yellow, and red polygons. The flight path information was represented by white/gray symbols linked by a magenta line visible in both views. The airplane information was represented by a 3D model in the 3D view and by an airplane symbol in the 2D view; both symbols were colored in orange. A compass was included to provide information on the direction the airplane was pointing toward in the 3D view.

The 3D view synchronized with the 2D view when the pilot selected an area of interest in the 2D view (i.e., the 3D viewpoint was adjusted accordingly). The user also had the option to adjust the size of the 2D or the 3D view dynamically by holding and vertically displacing the enhanced white view splitter in the middle of the screen. Weather forecast information was available when the "Play" button was pushed. A location button was implemented to allow pilots to center both views on the airplane's current position, as needed. Finally, basic iPad-based interaction controls (e.g., zoom in/out, rotate, pan) were available so that pilots could navigate both views for their own comfort.

I spent some time describing OWSAS so that you would understand its complexity from a technical perspective (all the experience and knowledge from expert pilots embodied in OWSAS must be accounted for as well). In addition, the following elements suggested that OWSAS' maturity (e.g., level of stability, reliability, trustworthiness) needs to be improved before we can put it to the test in a real flight situation: (1) its novelty in the context of commercial aviation; (2) the early-stage nature of the research; and (3) flying is life-critical (Millot, 2014). Not accounting for at least one of these elements and trying to test OWSAS in a real flight situation without discovering potential emerging properties (i.e., unexpected characteristics due to the integration of this new system in context that can have a positive or negative impacts on pilots' activities) would have been severe risk-taking in my opinion, because it can alter the safety of a flight. I needed to find a workaround to avoid such unnecessary risk-taking and therefore I decided to use the flight simulator (Figure 12.9) to perform HITLS with expert pilots. The more realistic the research study, the closer to real the insights

Figure 12.9 B737 flight simulator used for HITLS.

obtained. For this reason, it was important to have a realistic flight simulator with a high level of tangibility (i.e., flight cockpit including physical controls such as knobs, buttons, switches), to call on expert pilots that had a background with airplanes that were like the flight simulator (i.e., familiar with Boeing 737 series), and to design scenarios that were similar to what pilots usually experience during a flight. By doing so, I eliminated any risk-taking in a real flight scenario by navigating risks in the design phase.

12.5 PROPOSAL OF A MODEL FOR RISK-TAKING AND RISK-MANAGEMENT IN HCD

In the previous section, I mentioned a few key moments that I believe involved risk-taking and/or risk-management at various levels. In this section, I want to take the time to reflect on each key moment and to review the factors that influenced our final decisions. These reflections led to the proposal of a risk-management model specific to HCD for life-critical systems (Figure 12.10). This model is not a set of sequential steps to follow, but rather a set of elements to consider and reflect on when dealing with a multidisciplinary human-centered project. The model is composed of two areas: (1) a personal area that includes elements that relate to me, as a project leader or principal investigator; and (2) an external area that includes elements that I interacted with during this work. Each element is paired with one or more questions that are worth thinking about. Although these questions seem high-level, I believe they will make sense in the context of any similar future project.

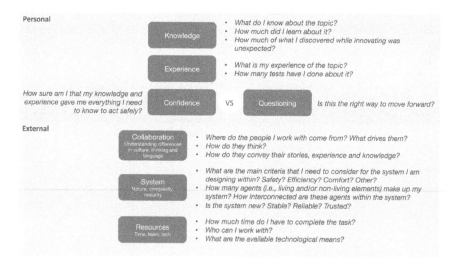

Figure 12.10 Proposed design-specific risk-taking and risk-management model.

12.5.1 Key moment 1 – interviewing pilots in flight conditions

The first key moment I described related to an experience of something that we had not done before. We decided to interview expert pilots as they operated a flight simulator, because we thought it would provide more context and allow them to generate a better perspective of how they deal with weather during a flight – based on environmental cues and the associated elicited memories. We observed that the first pilot we interviewed was cognitively preoccupied with performing a safe flight even though it was a simulation. We also realized that this involvement prevented him from answering our questions fully.

12.5.1.1 Factors

- **Exploration.** We decided to proceed with this methodology spontaneously without consulting the literature, adopting an exploratory mindset.
- **Knowledge.** We did not realize at the time that the realistic environment we put the pilots in would have such an impact on their behavior.

12.5.1.2 Learning

This lack of knowledge and experience led to insights contributing to our knowledge of the domain and the applicable methodology.

12.5.2 Key moment 2 – comparing two pilots' experience and knowledge

We decided to compare the opinions of two expert pilots with a view to pre-selecting the optimal solution(s) to represent 3D weather information. We made this decision because we had seen from experience that getting experts into the same room could open up debates leading to the emergence of meaningful insights. Rather than observing the debate in the context of our goal, we observed the interactions between these experts and myself during the discussion of the impact of 3D weather representations on weather radar attenuation.

12.5.2.1 Factors

- **Experience vs. exploration.** From Edwige's research and experience, we knew that this type of methodology could provide better quality insights, especially when working with pilots who do not know each other. However, we decided to perform this activity with two pilots that knew each other through flight and personal experience. Because of this scenario, pilot #2 talked significantly less that pilot #1, at least during the first phase of the discussion. The professional and personal relationship limited the quality of the debates as the two pilots tended to agree with many of each other's points. However, it allowed me to clarify a point because pilot #2 happened to understand me, and also understood pilot #1's mental model, he was therefore able to convey the point for me and serve as an intermediary.

- **Knowledge.** During the discussion, my lack of knowledge and understanding at the time impacted the quality of the conversation because I used technical and design-related terms that were not relevant to the pilots. This incompatibility was mainly due to our differences in work cultures (i.e., designer vs. pilot). It therefore took some time for the three of us to understand each other.

12.5.2.2 Learning

The lack of alignment in terms of language and work culture allowed me to observe how pilots and designers use distinct techniques to convey their respective points, and ultimately to find a common understanding. This meeting suggested that it might be worth giving control to a person familiar with both worlds to ensure this common understanding; this person would then become the mediator for the conversation (Boy, 2012).

12.5.3 Key moment 3 – final prototype evaluation: modifying the pilots' current model for interaction with the radar to allow comparison with OWSAS/designing the scenario without including the expertise of a pilot

Designing my last experiment was challenging because I faced a dilemma: on the one hand, I wanted to compare how OWSAS and the simulator's weather radar would support pilots' situation awareness and decision-making in a particular weather scenario. On the other hand, the integration of this specific weather scenario was not possible without modifying the simulator's weather radar. I made the decision to design an iPad-based replica of the radar and adapted the controls in consequence. A few pilots preferred this new layout because all controls were in one place right below the radar.

12.5.3.1 Factors (constraints)

- **Time.** My research scope was reworked and restricted due to limited funding toward the end. I therefore had to set up my last experiments with a lower level of maturity for OWSAS than I had anticipated. Because the system could not yet retrieve weather information from the simulator, I had to use a static weather scenario and the design of the iPad-based replica.

- **Anticipation.** The time constraint allowed me to anticipate and decide what path I should take when performing these experiments.
- **Knowledge.** My knowledge of potential tangibility issues (i.e., going from physical controls of the radar to updated iPad-based controls such as sliders and buttons) informed the risk involved when designing this replica.

12.5.3.2 Learning

- **Confidence vs. questioning.** I took weather classes to gain a more comprehensive understanding of what I was dealing with. I also continuously talked to expert pilots enrolled on the same program as me. I performed 15 in-depth interviews with expert pilots. I believed I had enough knowledge and did not take the time to question whether I should call on an expert pilot when designing my experiment.
- Taking the risk of designing a replica with potential tangibility issues revealed that the design adopted had advantages over the radar design in the simulator (i.e., all controls were in one place).

12.5.4 Key moment 4 – current research accounts for pilots only – the reality is different

In key moment 4, I explained that dealing with aviation weather is not only about pilots and the systems that they might have to support them. Adopting a HCD approach requires consideration of the whole ecosystem, because each element will affect how people perform their roles and jobs, how they use support systems, and how they communicate with each other in the ecosystem. The AUTOS pyramid revealed the potential complexity of this ecosystem. However, my research supervisor and I decided to focus on pilots first.

12.5.4.1 Factors

- **Experience vs. numbers.** In my research, I used a qualitative approach (i.e., gain a deep understanding of pilots' experiences vs. a deep understanding of general facts about them) which requires considerable time due to the nature of the tools used (i.e., one-on-one in-depth interviews, qualitative analysis, etc.). Because of this choice, I had to make sacrifices in terms of the scope of the project.
- **Complexity (scope) vs. resources.** Using a HCD approach revealed the complexity of the ecosystem we were contributing to, represented in the AUTOS pyramid. Although the best way to study a complex system is to account for all elements at once, I had to make sacrifices due

to time and resource constraints and decided to focus on the study of pilots. At the same time, I acknowledged the issue and suggested that one of the next steps in this research should be to account for air traffic controllers and dispatchers to understand the whole picture more accurately.

12.5.4.2 (Missed) learning

It is safe to say that I consciously missed the opportunity to collect insights that would have emerged from observing interactions between pilots and the other organizational agents. This realization is part of the conclusion to my design project. However, I hope that someone will take over the research and figure it out.

12.5.5 Key moment 5 – performing early-stage research for life-critical systems

Finally, I explained that it is a challenge to improve the maturity of OWSAS because it is part of a life-critical system. First, I detailed the elements that reveal OWSAS' complexity; I then listed other factors (e.g., novelty of research, low maturity of OWSAS) that added to the equation making it almost impossible to test it in real conditions. I then explained that using HITLS combined with high-fidelity elements (i.e., expert pilots, flight simulator, scenarios) was the workaround I chose to face this challenge.

12.5.5.1 Factors

- **Nature of the system being designed for.** The aerospace system and its components (e.g., pilots, air traffic controllers, dispatchers, airplanes, weather, etc.) are life-critical.
- **Complexity.** The current state of OWSAS includes several elements of varying natures and origins (i.e., content, visualization, and interaction concepts because of a mix of technical elements derived from the knowledge and experience communicated by expert pilots). The number of elements and their interconnections makes OWSAS a complex system. It is therefore very important to make sure that design for safety accounts for stability, reliability, and trustworthiness.
- **Novelty.** The research I conducted was relatively novel because contextualized insights (e.g., aviation weather in 3D evaluated by expert pilots in a realistic context – a flight simulator) were not readily available and needed to be generated on the go. My research was mainly generative as the literature contained very few similar studies that could be directly built upon.
- **Maturity.** Because my research was mainly generative, OWSAS was not mature and needed to grow from several points of view.

- **Safety.** Life-critical systems and their sub-systems are often complex, as demonstrated through the OWSAS project in the previous sections. It is crucial to account for safety as it could impact human lives.
- **Maturity vs. safety.** The nature of the system for which the design is destined, its complexity, and the novelty of the research led to a need to increase the maturity of OWSAS while also considering safety. HITLs were chosen as a workaround to address both concerns.

12.5.5.2 Learning

It is impossible to design a perfect system before it is actually engineered and launched in its real context, because simulation tools will always include imperfections. However, simulation remains crucial as it minimizes the possibility of safety issues emerging.

12.6 DISCUSSION AND CONCLUSION

This chapter describes a way of approaching risk that contrasts with traditional risk analysis (Hertz & Thomas, 1983; Zhang, 2011), where the risk of short-term accidents (Sadeghi, Dantan, Siadat & Marsot, 2016) is separated from the risk of long-term health complications (Neumann, 2004). By using HCD supported by Human Activity Development theory, we were able to understand operational risk at the design stage by performing a holistic and complex analysis of the end task, through feedback based on the expertise and experience of pilots. This approach gave us access to the details and nuances of the activities involved in a flight, which are indirectly linked to health and safety effects. In other words, we started learning about the pilots' activities through interviews, HITLS and other exercises, and progressively discovered the risks entailed and how they are managed. These risks were the focal points that driving the design of OWSAS.

The model we introduced in the previous section suggests that conducting a HCD project involves dealing with people (pilots), organizations (aviation administrations, air traffic control, etc.), and technology (weather systems) in a single system (Boy, 2012). Indeed, the key moments we described showed that using the HCD approach to design a life-critical system primarily requires group work.

To solve current issues in a specific context, it is important to understand the context and the people involved. Collaborating with expert pilots throughout the design project, using various methods and stimuli, allowed us to better understand (1) the world of aviation, (2) how pilots think and behave, and (3) the different challenges and issues faced by pilots while performing their job. Over the course of the study, we learned that it is crucial to make sure that one's understanding of a situation truly reflects the reality. In key moment 2, pilot #4 started acting as a mediator, making sure that both designers and pilots were aligned on the topic being discussed, which helped to avoid misinterpretations. This allowed us – designers and psychologists – to move forward in the design project with the right insights, avoiding potential risks later on. We then realized how much work with health and safety specialists, particularly in the prevention of professional risks, can be the key to anticipating and, above all, building new and safer operational perspectives with experts.

In addition, collaborating with the right people in context makes it possible to discover emerging properties that can be fed back into designs to mature the system or service and its safety, efficiency, and comfort. Using evaluative research methods and HITLS, we discovered some of these emerging properties in a close-to-realistic context that drove the next design iterations of OWSAS and the associated design changes. By doing so, we identified risk-taking scenarios and accounted for them moving forward, thus advancing the maturity of the system.

This collective aspect was also present internally. Indeed, Edwige (the psychologist) was there to support this work with co-construction methods, along with other researchers who provided their research input and advice.

These different perspectives supported me in adjusting the methods we used based on the dynamics and context of the research. From a methodological perspective, flexibility was key. While progressing with this design project, several constraints forced us to modify how we conducted the research. Key moments 1, 3, 4, and 5 showed that whether the challenge comes from the research environment (e.g., pilots being distracted by our environment) or from sources external to the project (e.g., time, budget), decisions had to be made if we were to achieve our goals. Making these decisions increased the level of risk, albeit to a measured extent; in each case, we balanced the pros and cons for each modification to the research and adopted the most appropriate with respect to the overall goal of the design project. The quality of the outcomes still reached our expectations; risks were managed at design time.

Both group work and flexibility in research methods allowed us to incrementally develop a system (OWSAS) to support the work of expert pilots. In conclusion, HCD is all about making the user shine and allowing them to develop their activity safely, efficiently, and comfortably in real flight situations in the future. It is also about making sure that the system is mature enough so that it can be used in reality.

Approaching and anticipating risk during design, while understanding and altering the real pilot's activity, was a means for us to attempt to reduce operational risk-taking in the future.

REFERENCES

Bonnemain, A., Bonnefond, J.-Y., Fontes, F., & Clot, Y. (2019). Vygotsky and work: An activity clinic to change the organization. In P. F. Bendassoli (ed.), *Culture, work and psychology: Invitations to dialogue* (pp. 59–80). Charlotte, NC: Information Age Publishing.

Boulnois, S. (2018). *Human-centered design of a 3D-augmented strategic weather management system for the aviation community*. Doctoral dissertation. Melbourne: Florida Institute of Technology.

Boy, G. (2012). *Orchestrating human-centered design*. London: Springer Science & Business Media.

Boy, G. A. (ed.). (2017). *The handbook of human-machine interaction: A human-centered design approach*. Ashgate: CRC Press.

Clot, Y. (2009). Clinic of activity: The dialogue as instrument. Learning and expanding with activity theory. In A. Sannino, H. Daniels & K. D. Gutierrez (eds.), *Learning and expanding with activity theory* (pp. 286–302). Cambridge: Cambridge University Press. doi: 10.1017/CBO9780511809989.019.

Endsley, M. R. (1988). Design and evaluation for situation awareness enhancement. In *Proceedings of the human factors society annual meeting* (Vol. 32, No. 2, pp. 97–101). Los Angeles, CA: SAGE Publications.

Engeström, Y. (2000). Activity theory as a framework for analyzing and redesigning work. *Ergonomics*, 43(7), 960–974. doi: 10.1080/001401300409143.

Engeström, Y. (2005). *Developmental work research: Expanding activity theory into practice*. Berlin: Lehmanns Media.

Hertz, D. B. & Thomas, H. (1983). *Risk analysis and its application*. New York, NY: Wiley.

Leontiev, A. N. (1979/1981). The problem of activity in psychology. In J. V. Wertsch (ed.), *The concept of activity in Soviet psychology* (pp. 37–78). Armonk, NY: M. E. Sharpe. doi: 10.2753/RPO1061-040513024.

Millot, P. (ed.). (2014). *Risk management in life-critical systems*. Hoboken, NJ: John Wiley & Sons.

Neumann, W. P. (2004). *Production ergonomics: Identifying and managing risk in the design of high-performance work systems*. Doctoral dissertation. Lund, Sweden: Lund Technical University.

Quillerou, E. & Boulnois, S. (2020). Analysing human activity development through (work) spaces in a human-centered design project for commercial airline pilots. *Journal of Workplace Learning*. https://doi.org/10.1108/JWL-02-2020-0022.

Sadeghi, L., Dantan, J., Siadat, A., & Marsot, J. (2016). Design for human safety in manufacturing systems: Applications of design theories, methodologies, tools and techniques. *Journal of Engineering Design*, 27, 844–877.

Vygotsky, L. (1931/1978). *Mind in society: The development of higher psychological processes*. Cambridge: Harvard University Press.

Zhang, H. (2011). Two schools of risk analysis: A review of past research on project risk. *Project Management Journal*, 42(4), 5–18.

Chapter 13

Designing sustainable 'plastic' work systems

A resource for work-related prevention in France's waste management and recycling sector

Leïla Boudra
Conservatoire National des Arts et Métiers
Paris, France

Valérie Pueyo and Pascal Béguin
Université Lumière Lyon 2
Lyon, France

CONTENTS

DOI: 10.1201/9781003221609-13

13.1 INTRODUCTION

Health and safety at work are closely linked to how work systems are designed. For many years now, it has been understood that design contributes to preventing occupational risks. This can be achieved through integrating regulatory standards that aim to ensure workers' safety when they interact with machines. But it is another challenge to offer room for maneuver in workers' activity at the design stage. As defined in ergonomics, rooms for maneuver are resources for workers' activity (Caroly, Coutarel, Landry, & Mary-Cheray, 2010). It offers possibilities for workers to develop various ways of working (i) to reach production targets, (ii) to cope with anticipated and random variabilities, and (iii) to avoid adverse effects on their safety and health. Workers *"operate positively and meaningfully in relation to the context at hand, then it is important to expand the repertoire of resources they will have available to them in their work environment"* (Béguin, Owen, & Wackers, 2009, p.7). In that way, such resources for work activity are useful levers for prevention.

This way of occupational risks prevention is understood marks an additional step beyond the expectations laid down by the regulatory framework.[1] In France, the regulatory framework sets as targets to prevent or – failing that – to limit exposure to the identified risk factors to avoid their consequences on people (accidents, diseases, etc.). These goals reflect a defensive vision of health and safety at work, rooted in a hygienist tradition of occupational health policy. Without denying the relevance of such objectives, limiting prevention to these regulatory aspects could result in neglecting the active role played by workers in their work and in prevention of the associated occupational risks (Nascimento & Falzon, 2012; Pueyo & Zara-Meylan, 2014), as well as masking the links between work and health, including positive effects on the development of the workers. An efficient prevention must integrate the real work activity and the strategies workers develop to manage and cope with occupational risks and to preserve their health. Furthermore, failing to integrate the reality of work into prevention could lead to a "disconnection" between prevention actions and the industrial realities and industrial project of the company (Boudra, 2016).

Therefore, to prevent exposure to risk factors and to support the development of workers' health and resources for their activity, we propose to consider occupational risks prevention as a "work-related prevention" in line with the company's performance challenges, including the technical and organizational systems and the real work activity. This is a prevention "in connection" (*ibid.*), rooted in an "offensive" vision of the relationships between humans, their work, and their health.

On a pragmatic level, a work-related prevention in connection aims to create efficient conditions for workers to be fully involved as stakeholders in their work. In other words, it assumes that workers are provided with the conditions necessary to be "capable subjects" in their work situation, by using material and psychological resources to better adapt their activity to the reality, and through their actions, capable of transforming the reality and themselves (Rabardel & Pastré, 2005). "Activity" is to be understood – in the ergonomic sense – as a set of actions that are finalized (object-oriented), unique (a given subject, a given context), and carried out by an individual in interaction with the other components of the system: the environment, technical objects, and other people (Carayon, 2006; Daniellou, 2005).

From this anthropocentric perspective of work-related prevention, the technical systems are considered as resources for activity (Rabardel & Béguin, 2005). These issues raise questions about an instrumental approach of design and prevention, oriented through the simultaneous development of artifacts, situations, and individuals (Béguin & Rabardel, 2000). Design should therefore incorporate users' activities and the organizational conditions that will, in part, determine the uses *in situ*, i.e., the ways in which the artifacts are embodied in the users' activities (Béguin, 2007a; Folcher, 2003).

However, the possible organizational forms are closely linked to the characteristics of the technical systems. Once the technical and technological choices have been defined, any forms of organization (even those that might be desirable and/or desired) can no longer be implemented. Consequently, it is necessary to consider simultaneously all the dimensions of any work system: technical, organizational, and user-activity-related.

So, how can we design work systems that offer as broad a range of organizational forms as possible, and that incorporate efficient room for maneuver to make workers "capable subjects"? In this text, we develop an important dimension of work systems that we call "technical and organizational plasticity" (Boudra, *op.cit.*). "Plasticity", in the sense used by Béguin (2007b, p.118), means the property of a system that *"leaves the activity sufficient freedom for maneuver to render technical aspects more efficient whilst remaining in good health"*.

Our contribution is based on a research in ergonomics, funded and performed in partnership with the *Institut National de Recherche et de Sécurité* (INRS, France's research and safety institute for the prevention of occupational accidents and diseases). The research aimed to develop work-related prevention in the sector of waste sorting and recycling in France, when designing future material recovery facilities (MRFs). From our point of view, as MRFs contribute to sustainable development, prevention and design must be oriented through sustainable work systems (Boudra, 2016). In other words, three dimensions need be considered simultaneously:

- Economic, notably the efficiency of the production systems.
- Environmental, notably the increase in waste recycling and the commercial quality of the outgoing products.
- Social, notably the work sustainability (in the sense used by Docherty, Forslin, & Shani, 2002), i.e., work that does not degrade health but rather develops the capacities of the individuals.

In this chapter, we will first describe the context of the research, the corpus, and the method chosen to analyze the sorting work activity and the industrial performance challenges for sustainability. Then, we will focus on designing work systems that are "plastic" in terms of both their technical and organizational dimensions. Finally, we will go back over the lessons that can be learnt for a work-related prevention that is "in connection".

13.2 CONTEXT OF THE RESEARCH

MRFs in France are industrial firms that contribute to sustainable development and its three pillars: economic, environmental, and social (Pearce, 1988). These firms receive recyclable household waste. The waste is then sorted according to type of material, before being packaged and stored. MRFs are process industries that combine manual activities with equipment resulting from innovative technologies. Distinctions can be made between three categories of sorting:

- Automated sorting performed by sorting machines, for example infrared systems that detect and separate products based on their material composition.
- Mechanized sorting performed by machines separating products according to their ballistic density.
- Manual sorting performed by employees around a sorting belt and aiming to achieve more precise sorting by material and to correct machine errors.

Manual sorting constitutes the core of the economic and industrial activity of the sector, and it is an essential step in any industrial process. But the working conditions of waste sorting workers raise concerns about the risk factors for health and safety relating to:

- Exposure to the waste, one of the most studied aspects that can be a risk factor notably for biological and respiratory diseases related to the presence of molds, bacteria, dust, or soiling (e.g., Malmros, Sigsgaard, & Bach, 1992; Poulsen et al., 1995; Rushton, 2003; Schlosser et al., 2015).
- Work organization and work environment (shift work done on an automated line) that can be a risk factor for accidents and physical injuries, work-related musculoskeletal disorders (WRMSDs), pathologies related to postural constraints, and risks of psychosocial disorders (e.g., Engkvist, 2010; INRS, 2011; Lavoie & Alie, 1997).

Developing a work-related prevention approach and improving working conditions are thus major issues for this sector that is undergoing industrialization and whose choices as regards technical processes are increasingly focusing on combining automated sorting with manual sorting.

13.3 PRESENTATION OF THE FIELD OF THE RESEARCH AND METHOD

The research implied five MRFs in France, all sites volunteered to take part to the research:

- One MRF (referred to as "MRF1" in Figure 13.1) for an exploratory study, conducted over six months. The target was to develop knowledge on work activity in this sector, and to better understand health and safety issues for other MRFs.
- Four MRFs (referred to as "MRF2", MRF3", "MRF4", and "MRF5") for analyses over 24 months.

These five MRFs belonged to public structures owned and run jointly by several local districts (*communes*). These public structures are known in a generic way as "public inter-*commune* cooperation establishments" (or EPCIs according to the French abbreviation[2]). These ECPIs own the production facilities and contract out to private groups for the industrial operation of the sites. The tasks of the operating group, represented by the director of the sorting site, were to manage production, and to recruit and manage staff.

Figure 13.1 Picture of the mobile disposal equipment used to store the large-sized ferrous waste, manually removed from the belt (MRF4).

Table 13.1 Main characteristics of the five material recovery facilities

	Date commissioned	Annual processing capacity (in metric tons)
MRF1	2000	10,000
MRF2	1998	4,500
MRF3	2006	10,000
MRF4	2008	10,000
MRF5	2011	32,000

These five plants were equipped with distinct technical systems and had also different annual processing capacities: from 4,500 metric tons per annum for the smallest, to 32,000 metric tons for the largest (Table 13.1). Such diversity in annual processing capacities denotes the increasing industrialization of the sector and is accompanied by increasing levels of mechanized and automated sorting.

The analysis method proposed came under activity-oriented ergonomics (de Keyser, 1991; Wisner, 1995). More precisely, three steps can be distinguished.

The first step aimed to produce a work analysis, i.e., creating knowledge about work activity and about the conditions under which the work is carried out (prescribed work and real work, determinants, and effects on workers' health and safety, and on the system). This analysis took place over 24 months and concerned the sorting workers, their managers, and officials from the EPCI, who were assigned to monitor administrative and technical operations. We performed 22 analysis campaigns, lasting two days each on average, during which 293 video sequences were recorded. Then we realized 38 auto- and allo-confrontation interviews[3] (Mollo & Falzon, 2004), followed by meetings with the employees to validate and expand our work hypotheses with them. Each analysis campaign was systematically supplemented by an average of four individual interviews with the team managers and production managers. Finally, semi-structured interviews were conducted with the officials from the EPCI.

The second step, that we call "objectification of work", concerned only four sites (excluding the first exploratory one). Meetings were organized with the sorting employees and their team managers, during which the findings from the analysis performed in the first step were presented and discussed for collective validation and greater understanding. Two meetings were held on each site, the first meeting was held with the sorting employees and their team managers, and the second one with the decision-makers, i.e., the team managers and the production managers of the MRF and the officials of the EPCI, at the level of the local area or "territory" covered by the MRF.

The third step pursued a goal of transformation. Based on the results discussed and validated during the meetings at step two, the objective was to focus on how to develop a dynamic prevention process in all MRFs, particularly when designing future MRFs and transforming work situations.

13.4 FINDINGS OF THE ANALYSIS OF THE WORK ACTIVITY IN MATERIAL RECOVERY FACILITIES

13.4.1 The work activity of the sorting employees and its determinants

A typical sorting process is subdivided into three spaces grouping together several workstations within one or more sorting cabins,[4] organized around one or more automated conveyor belts.

- First: a "presorting" space where bulky waste (e.g., large cardboard boxes), the main nonrecyclable household waste, and dangerous objects (gas bottles, metal objects, glass, etc.) are removed.
- Second: a "fibrous waste sorting" space for paper and cardboard.
- Third: a "nonfibrous waste sorting" space for plastic (e.g., bottles, or containers), aluminum or steel packaging (e.g., cans or aerosol canisters), and food and beverage cartons.

Sorting employees work at the various stations, shifting according to the schedule planned by managers. At each workstation, they must remove an average of three types of recyclable waste (cardboard, paper, plastic packaging, etc.) or nonrecyclable waste (undesirables, soiled products, etc.) from the belt and to throw them into the disposal equipment provided for that purpose. Some items of disposal equipment are fixed, i.e., they were included at the design stage and cannot be moved because they are connected directly to the space where the sorted waste is stored. Others are mobile or moveable (such as containers, crates, bins, etc.), i.e., they have been added to collect additional waste at the various workstations. The items of disposal equipment are located on the sides, backs, or fronts of the workstations.

Production is assessed by the production managers in terms of quantity (metric tons per hour and productivity per sorting worker), and by the recyclers in terms of quality after reception of the sorted waste. The recyclers set a minimum quality level to be attained, determined based on the material sorted. For example, a bale of cardboard packaging should be 97% cardboard waste; 3% other materials are tolerated by the recycler.

To reach the production quantity and quality targets, our analyses of the work activity revealed that workers deployed numerous adaptive strategies both individually and collectively. Those strategies result from choices and trade-offs made individually and collectively by the workers, based on their experience (Pueyo, Toupin, & Volkoff, 2011; Sperandio, 1971;

Teiger, 2007; Wisner, *op.cit.*). Production has multiple variabilities that are more or less predictable and controllable, e.g., (i) regarding the types of waste related to the products consumed and the quality of the sorting performed by householders, (ii) regarding how the waste is collected and transported to the MRF, and (iii) regarding how waste is stored in the plant and whether it is processed by automated and/or mechanized equipment (and therefore how well the machines operate), etc. (e.g., Boudra et al., 2019.)

More precisely, the results of the work activity analyses show three categories of adaptive strategies used in the MRFs:

- Individual adaptive strategies, at workstation scale, relating to coping with the quantity of waste to be sorted in view of the quality, identified by sight, touch, or smell.
- Collective adaptive strategies that mainly relate to mutual assistance between workers on the same sorting belt. And other collective strategies are implemented by employees who load the waste upstream of sorting. Those workers deploy strategies while loading waste onto the conveyors bringing it to the sorting cabin, with the aim of smoothing out the flows of waste, thereby avoiding machine incidents (such as jams) and facilitating work for the employees sorting the waste in the cabin.
- Organizational adaptive strategies, put in place by team managers and related to adjustment of production (quantity, belt speed, etc.) and to assignment to workstations, e.g., moving workers from one sorting workstation to another, depending on the volumes and quality of waste to be sorted.

In nominal mode, such adaptive strategies are essential to keep up with the industrial performance criteria while also preserving health, as the following examples show.

On a collective level, the adaptive strategy most frequently used is mutual assistance between workers on the same sorting belt. It consists of helping a person to do his or her work or performing an action in his or her place (Avila-Assunção, 1998). Mutual assistance is particularly significant in production line work, as the workstations are dependent from the start to the end of the chain. As a female worker of MRF1 explained:

> We try to help our colleagues, and as far as I'm concerned, we work in collaboration. If waste is not spread out [upstream], if we don't help one another, the work is difficult, and we don't meet the targets.

On an individual level, this worker deployed adaptive strategies aiming to regulate production while preserving industrial performance and her health. During an auto-confrontation interview, she described one of her strategies using gestures that she commented verbally[5]: by using her arm as a filter, the movement consisted in blocking the flow of waste downstream to sort it upstream, while at the same time allowing a certain amount of waste to pass through. This strategy had three aims: (i) to deliver performance for the sorting carried out (in terms of quality), (ii) to give the worker more time to perform sorting by circumventing the throughput rates imposed by the automated belts, and (iii) to avoid overloading subsequent stations and regulate the flows of waste. Adaptive strategies thus offer workers additional room for maneuver to attain quality targets and to limit consequences on their health.

These initial findings of the activity analyses show the importance of adaptive strategies for meeting the industrial performance challenges of sorting companies. In a production line work context, characterized by a dependency of workstations along a production line, workers play a fundamental role in managing and regulating production. To do this, they must make trade-offs and choices. Production and quality issues are always central to such trade-offs and choices, but they are sometimes detrimental to the health of the workers, particularly in degraded configurations. This is what we will look at in the following part.

13.4.2 Work performed in degraded mode due to mismatches between technical system and the waste characteristics

In *degraded mode*, disruptions occur in work systems but do not necessarily and systematically require shutdown of the production systems. However, the persistence of these phenomena has consequences of greater or lesser significance on production and on working conditions (Kerval, 1990).

We studied many manifestations of this degraded mode during our research in MRFs, and we will present a representative example. Residents' consumption habits are a significant determinant in the composition of the waste collected. These habits evolve over time, depending on lifestyles, on modes of consumption, and on the products put on the market by manufacturers. Thus, in the MRFs, numerous waste items appear that have new characteristics, modifying their densities, their dimensions, or their material composition. This evolution of the waste characteristics is neither integrated nor anticipated at the MRF design stage. Consequently, mismatches tend to disrupt operation of the technical equipment or affect the working conditions for workers sorting waste manually. New adapted solutions must be found locally. However, due to investment costs or existing technical configurations, technical adaptations are not systematically possible.

An example can be developed from the case of MRF4, where large ferrous waste pieces were removed in the presorting space. Ferrous products were usually removed from the process by an overband magnetic separator.[6] But large-sized pieces could not be processed like other metal products, to avoid risks of technical malfunctions (jams, breakages, etc.). The workers positioned at the presorting belt were expected to manually remove such ferrous waste. These operations involved high biomechanical and load-carrying stresses, due to the weight of large-sized pieces, and thus exposed workers to a risk of injuries. A large-capacity container (660 L) had been placed inside the sorting cabin, near one of the workstations (Figure 13.1), to store the large ferrous waste removed. However, this container made it difficult for the employees to move around in the sorting cabin and would prevent them from evacuating the sorting cabin quickly in case of an emergency. The weight of this large container, once filled, required at least two employees to move it out of the sorting cabin and empty it.

MRF2 had faced similar problems regarding the diversity of waste. As in MRF4, employees at MRF2 were expected to remove large-sized ferrous waste manually, before waste passed under the overband separator in the center of the sorting belt, in addition to their other sorting tasks. However, our observations indicated that not all waste could be removed from the belt, given (i) the quantity and the diversity of the waste on the sorting belt and (ii) the speed at which the belt moved. When large-sized waste was picked up by the overband magnet, it caused jams in the metal press (a specific press made for steel waste located at the disposal equipment). These jams then had to be cleared by workers. Using an "unjamming" pole designed by the workers themselves, one of them poked inside the disposal equipment of the overband separator to clear the inlet of the metal press, blocked with a heap of steel waste (Figure 13.2). During the "unjamming" operation, the technical process was stopped by the employees themselves, until the operation was complete. These steps had an impact on production targets.

In the short term, this operation was associated with various accident risk factors for workers: injuries due to being hit with the pole, or related to the muscular efforts and to the biomechanical stresses required to perform this operation, or risks of projection of ferrous fragments leading to injury,

Figure 13.2 A worker performing manual clearance of a jam at the metal press in the sorting cabin (MRF2).

especially on the unprotected areas of the arms and the head, etc. In the longer term, such repeated efforts could also damage health, for example by inducing WRMSDs.

These two examples show what characterizes degraded mode operation in the MRFs and the consequences for the workers and their safety and health. When the production process operates in degraded mode, manual sorting is performed to correct mismatches between the technical system and the characteristics of the waste. Such mismatches occurred in all the MRFs studied. They are detrimental to the achievement of production targets and the malfunctions caused must be corrected by the workers. In addition to the costs in terms of production, there are costs to the health and safety of employees exposed to WRMSDs and accident risk factors. In the following part of this chapter, we will see how these mismatches are related to territorial waste management in disconnection from the technical systems designed.

13.4.3 Decoupling between the technical system designed and the territory-anchored waste

The characteristics of the incoming waste for MRFs depend on the local area or "territory" where they are produced: defined by the lifestyles and modes of consumption of the residents, economic flow, tourist flow, etc. (Boudra, Pueyo, & Béguin, 2018; Boudra et al., *op.cit.*). The analysis of the work activity indicates that the adaptive strategies developed by sorting workers are based on knowledge acquired through experience and referred to as "territorialized operational knowledge" (Boudra, *op.cit.*). The workers have thus developed criteria to identify the collection channels of the incoming waste during the sorting operation, or indeed sometimes the origin of a particular batch of waste within the collection channel. Territorialized operational knowledge involves identifying the types of waste items present, their frequency on the sorting belt, and the states they are in (damaged, deteriorated, soiled, or hazardous products, etc.). Territorialized operational knowledge includes indications that are useful for the action, constructed by workers from their own sensory information – sight, hearing, smell, and touch.[7]

More specifically, the waste collected for each MRF is dependent on three territorialized dimensions that we have characterized (*ibid.*):

1. The transport modes – waste is brought directly to the MRF by garbage trucks after the waste collection rounds or, for geographical areas that are the furthest away, goes via transfer centers[8] and then are delivered to the MRF.
2. The collection modes – in bags or individual and collective containers for door-to-door waste collection, or in bulk via voluntary drop-off points.

3. The composition of the collected waste relative to the sorting instructions given to the residents by the local authorities (packaging only or multi-materials[9]).

These three dimensions determine the territorialized waste collection channels. But those channels are not considered when designing the technical systems (*ibid.*). The design is based on the recycling channels, i.e., on the commercial logic supporting waste recycling. For example, a machine that mechanically separates cardboard boxes from other waste should be installed from a minimum annual production of recycled cardboard boxes, as it then becomes economically advantageous. But by reasoning in terms of overall volume, it is impossible to consider the diversity of waste currently processed in the MRF. In addition, this reasoning does not allow to identify how it would be suitable to install such a machine, the criteria of which are not based exclusively on cost-effectiveness (economic targets). Other criteria, such as the efficiency of waste separation (environmental targets) and the preservation of health and safety (social targets) should also be considered. Cardboard packaging can come in various forms, volumes, and states (some may have become wet during transport, for example). Depending on this diversity, such waste cannot be sorted in the same way either by mechanical or by manual means. Nevertheless, such diversity is incorporated and handled only in the finesse of the work activity. Furthermore, the waste characteristics that are related to these three dimensions do not remain constant over time. They change without it being possible to accurately predict the forms that any such changes will take.

Two dimensions are thus considered independently:

- On the one hand, technical systems are designed according to the recycling channels, in a context where recyclers are increasingly demanding in terms of the quality of raw materials produced by sorting, rather than being designed according to the territorialized channels of the catchment area. Therefore, the characteristics of the waste collected and stored in the MRF – the incoming collection channels that we presented above (see Section 13.3) – are not precisely specified upstream to guide the design choices.
- On the other hand, the territorialized waste collection channels reflect the political choices made in the territory and are not automatically associated with organizational or technical options within the MRF.

A *decoupling* is thus created between the characteristics of the territorialized waste and the possible technical and organizational forms of the MRF. The decoupling should be regarded as a state of tension between the two dimensions of the territory-anchored waste and the technical and organizational design of MRFs that have been considered independently and separately from each other.

This tension gives rise to mismatches and increases over time. And this increased decoupling stems simultaneously from changes in the territory and changes in the MRF:

1. The changes in the territory partially or totally modify the incoming waste collection channels. The territory is a heterogeneous administrative entity, with a set of stakeholders who have their own discretionary power as to the policies they implement, such as the waste sorting management (e.g., choices for collection methods: containers, bags, voluntary drop-off points, etc.). These policies are themselves susceptible to change (elections of representatives in local authorities and/or changes to national decentralization policies, emergence of new economic activities, etc.). Therefore, the territory appears as a constantly evolving entity: technically, socioeconomically, organizationally, and politically.

2. The capacity of the technical system to absorb changes appears to be limited. For existing sites, the need for maintenance becomes increasingly important as the technical systems "age", making it more difficult to absorb changes without extensive technical transformations. But at the same time, the increase in the automation of the MRF's technical systems makes them more and more vulnerable to changes, in consequence of the lack of consideration of the diversity of the waste treated and their territorialized anchorage. The already underspecified bases on which the systems are designed thus become no longer relevant or valid.

The more significant the decoupling, the more difficult and costly the adjustments will be for the workers. Faced with this decoupling, finding adaptive strategies to manage these tensions becomes a crucial part of workers' activity. In fact, production performance relies on the ability of workers to implement adaptive strategies in their activity to cope with the inconsistencies generated that increase over time. It is then incumbent on workers to resolve these tensions, by individually and collectively deploying adaptive strategies to achieve the required production targets.

Consequently, improving work-related prevention in MRFs requires the decoupling between a given MRF and its territory to be limited. Work-related prevention can only be developed if the design and the possibility of adapting equipment to absorb territorial, organizational, and technical changes are considered. But the issues are not exclusively technical. It is also essential for decision-makers to integrate the technical possibilities of the MRF before considering changes to be made in the territory.

Considering these two aspects is a challenge when seeking to develop sustainable work systems, not only as it limits the consequences on health and safety at work but also as it develops the potential for waste recycling by reaching the industrial and commercial targets. These are precisely the objectives of a sustainable development strategy (e.g., economic, environmental, and social pillars).

How, then, can we foster a sustainable coupling between the MRF and its territory? In this research, our answer was to work with decision-makers in waste sorting inside and outside the company to steer the future technical choices toward "plastic" work systems, as we will present in the following section.

13.5 DESIGN "PLASTIC" SYSTEMS AS RESOURCES FOR WORK-RELATED PREVENTION

13.5.1 Define "plasticity" for work-related prevention

We have pointed out that workers' adaptive strategies were essential in MRFs to resolve the tensions generated by the decoupling between the MRF and its territory and then to achieve the required production targets. The various adaptive strategies implemented represent many possibilities for the workers individually and collectively to deliver productive efficiency and sorting quality while also considering their health, i.e., to make the sorting systems more efficient. But the possibilities of deploying these strategies and their degree of efficiency are closely linked to the organizational and technical forms of the site where they work.

The technical system is one possible form that is specified during the design stage based on the state of the available knowledge and of the information provided by the EPCI that owns the site and is also the commissioning client. That form is linked to the political, socioeconomic, local, and historical dimensions of a given territory in that it embodies or crystallizes a set of representations related to the territorial characteristics of the waste at a given time – the time at which the design is decided. Those representations are then translated into a chosen technical system. How then is it possible to consider changes in the territory and in the technical system when designing the work systems to achieve smooth coupling?

Maintaining a "sustainable" coupling between the MRF and the territory can be viewed through the prism of technical-organizational "plasticity". In general, the concept of "plasticity" refers to a characteristic of a system that can allow sufficient margin for activity to ensure the system's efficiency and guarantee the workers' health (Béguin, 2007b). Thus, plasticity enables workers "to adapt or negotiate their own individual or collective courses of action, to reshape action and to draw on resources as needed" (Owen, 2009, p.101). The work systems' plasticity is thus a non-negligible focus for developing new room for maneuver, useful for improving work-related prevention approaches. Moreover, for prevention "in connection", such as the approach we propose, plasticity is a lever for (i) considering the development of individuals, of their health, and of socio-productive systems and (ii) contributing to the design of sustainable work systems.

To our knowledge, plasticity has mostly been studied through the lens of technical object properties (Béguin, 2007b, 2010). The main idea is that the artifacts are used in multiple contexts of action. Considering *in situ* practices with the participation of users provides resources to design artifacts that are sufficiently adaptable to make them efficient and to offer possibilities of room for maneuver in workers' activity.

However, this assumes that work and production are organized in a sufficiently flexible and plastic mode, i.e., in a mode that can adapt to changes and to fluctuating strategies. Once the technical aspects are crystallized, all forms of organization, even if they are desired, are no longer possible. Consequently, the challenge is to implement work and production organizations in which the workers can build and deploy suitable responses to the problems arising in real time as they work. Therefore, work systems need to move toward "technical-organizational plasticity", allowing workers the possibility to adapt their activity to local circumstances (Vicente, 1999), without endangering their health.

From the work point of view, plasticity therefore depends on the room for maneuver that give the individuals the possibilities to cope with the contingencies of the situation, while also remaining in good health. Any work situation is determined by technical and organizational "boundaries" (Carayon, *op.cit.*; Vicente, *op.cit.*) that define a "prescribed" framework within which workers can deploy their activity and deploy adaptive strategies. Thus, the range of possible types of room for maneuver is directly linked to the organizational and technical dimensions of the work system. Indeed, those dimensions define the boundaries for carrying out an action, within which adjustments are possible, but beyond which production efficiency deteriorates (Vicente, *op.cit.*). If the framework is too restrictive, the possibilities for the workers to deploy adaptive strategies are then also restricted and make it very difficult to deal with the industrial, organizational, and intra-individual variabilities. Health problems can then arise for the employees (WRMSDs, psychosocial risks [PSRs], etc.) and at the same time difficulties for the company in reaching production targets. It is thus easy to understand the advantage for work-related prevention of the design of sufficiently flexible technical and organizational boundaries.

Technical-organizational plasticity thus suggests considering both the technical artifacts and the environment in which it is to be achieved, based on the users' activity. More precisely, what can be described as the plastic properties of a technical and organizational system are related to the anticipation of the possible forms of organization and activity of workers for efficient actions (Daniellou, Escouteloup, & Beaujouan, 2011; Owen, 2009). Anticipation is not a prediction of the activity (Vicente, *op.cit.*), which would be impossible since many components of human activity cannot be foreseen. Rather, activity is characterized by its situated and singular dimensions (Suchman, 1987). Anticipating such possible forms is a way of defining a "space of functional action possibilities" where the workers are going to operate (Daniellou et al., *op.cit.*). In other words, it constitutes a resource for employees' activities that lets possibilities for them to develop the range of adaptive strategies favorable to health (Béguin, Owen, & Wackers, 2009).

In the following section, we will identify the technical and organizational dimensions of plasticity that we proposed for the MRFs.

13.5.2 Characterize the forms of plasticity

Technical-organizational plasticity of MRFs can be specified based on a detailed analysis of the activities performed by the workers. This analysis that we carried out in steps one and two of our methodology (see Section 13.2), contributes to the identification of the possible forms of activity in a defined environment of use. In the MRFs, the idea is therefore to consider the work system as a unit anchored in its territory and made up of technical objects (mechanized sorting, automated sorting, etc.; see Section 13.3), and for which the trade-offs and choices are linked to the activity of multiple individuals acting individually and collectively in defined organizational forms.

We proposed three recommendations to support the technical and organizational plasticity in MRFs, concerning (i) the sorting workspace and the team management, (ii) the adaptation of the technical equipment and the work organization, and (iii) the adjustments of the machines to the type of waste to be sorted. All these dimensions foster the development of the individuals, the groups, and the organizational resources to give workers more freedom to build strategies that allow them to protect their health while reaching their production targets.

The first recommendation aims to promote the development of collective resources by bringing together the sorting employees and organizing the sorting work in a single workspace. Currently, sorting cabins are designed as a single workspace or are subdivided into several distinct spaces. The workspace organization is in fact a mirror of the choices of production organization.

For the MRFs where we performed our research, only one – MRF2 – was designed with a single sorting space. It was the oldest MRF, the less mechanized, and was not equipped with any automated equipment. Only one sorting belt was present on the site, for all operations (presorting at the beginning and then sorting of nonfibrous and fibrous waste).

Subdividing the sorting cabins into multiple spaces corresponds to a form of subdivision of the production (into presorting, sorting of fibrous and nonfibrous waste), associated with technical equipment for mechanized and automated sorting, the function of which is to separate the waste into various flows corresponding to steps in the sorting process. For example, in MRF1, sorting was divided between three sorting cabins. The first cabin was the presorting cabin, corresponding to the first step in the process. After the presorting step, a machine separates the fibrous waste from the nonfibrous waste to form two flows, each of which is conveyed on a distinct conveyor belt, one leading to a cabin where nonfibrous waste is sorted; and the other leading to a cabin where fibrous waste is sorted.

In the case of MRF3, its initial design included a single sorting cabin in which two sorting conveyors were located. When new automated sorting technology was introduced, one of the belts was removed and two new workstations were designed in a confined space adjoining the sorting cabin. Refitting the sorting cabin would have been too expensive and would have required production to be totally shut down during the operation period.

Thus, the introduction of new technical and automated forms in MRFs led to a subdivided workspace. However, that form of production and work organization does not seem to be favorable to collective adaptive strategies. Conversely, when the workstations are grouped together into a single space, multiple interactions in the collective work emerge and make it possible to deploy adaptive strategies to regulate production. Such interactions contribute to the efficiency of the system while also managing the health and safety risks (Marc & Rogalski, 2009).

Collective adaptive strategies are also promoted by the presence of team managers in the sorting cabin, where they encourage and organize collective strategies. In addition, they set up what we have called adaptive organizational strategies, adjusting the organization in line with variations in production. For example, managers adjusted in MRF5, by moving an employee from one workstation to another based on difficulties identified in real time, or in MRF3, by reducing the metric tons processed per hour and adjusting the speeds of the sorting conveyors according to the quality and quantity of the corresponding waste flow.

The second recommendation focuses on equipment at the workstation. The single sorting workspace must also have equipment adapted to the activity of workers in the technical configurations of the sorting cabin and the workstations.

For example, mobile disposal equipment is essential for the activity because changes in materials and in modes of consumption by residents cannot be anticipated accurately at the design stage. Mobile disposal equipment exists in different volumes and forms (containers from 120 to 660 L, crates, bins, etc.) due to the need to adapt the container to its contents. Large cans can be contained only in sufficiently large mobile disposal equipment (see example for MRF4 Section 13.4.2). Similarly, small plastic objects, such as yogurt pots or meat-packing trays fly easily due to their low ballistic density. They must be stored in fixed or mobile containers located in the immediate vicinity of the workstation, otherwise they would simply end up on the floor.

Such diversity in the available mobile disposal equipment invariably works in favor of plasticity. Furthermore, it is necessary to have adequate supplies of mobile disposal equipment for every workstation, and to be able to place the disposal equipment around the workstation to meet the workers' requirements. This arrangement around the workstation should not limit the possibility to move along the walkways of the cabin to move and empty the mobile disposal equipment as easily as possible. Insufficient space to move the mobile container and access the fixed disposal equipment (where the waste collected in the mobile equipment is emptied) would require workers to adopt awkward postures in addition to carrying heavy loads when handling the containers.

From a technical point of view, providing adapted mobile equipment is a first step, but it must be accompanied by organizational measures to promote plasticity within the system. It is therefore also essential to provide an organization for emptying mobile containers. In fact, organizing the time for emptying of containers is required, first to limit the load carried individually and second so that this task is not performed "in the background", i.e., without being accounted for in the process time. During our analysis, we observed that, in many MRFs, the mobile disposal equipment was emptied by the employees during their breaks, i.e., exceeding working time. The organization should also aim to propose that the mobile disposal equipment be emptied more regularly to limit the amount of waste overflowing from it that regularly clutters floors and walkways, so as to avoid falls due to tripping. The organization should also focus on proposing more regular emptying of mobile equipment to limit the amount of waste that regularly impedes circulation over floors and walkways, to avoid accidents related to falls.

The third recommendation relates to the capacities of the various machines to adapt to the territorialized waste. As presented above (cf. Section 13.3), waste sorting is characterized by a high industrial variability, specifically related to the territorialized dimensions of the waste. However, those dimensions are not considered during the design of the work systems. One reason for this may be a lack of knowledge of the territorialized collection channels, or non-anticipation of changes to the decision-makers (funding providers, production heads and executives, work organizers, etc.) by the technical designers. Current uses and future needs usually remain under-specified (Visser, 2007), and consequently the technical and organizational solutions adopted at the time of design choices are not sufficiently plastic to be adapted later.

Let us take the example of MRF5. This site was equipped with machines to separate fibrous from nonfibrous waste, and with an automated machine to sort plastic waste. Soon after it was opened, the site experienced quality problems that the production managers attributed to a problem with the mechanical separators. Despite some efforts to solve the problem, no real improvement was achieved, because the mechanical ballistic separators of the machines could not be inclined, and therefore could not be adjusted according to the various collection channels and the variabilities of the waste. Our analyses showed that for multi-material collection channels, the employees at the fibrous waste sorting belt worked at an average rate of 1.85 items of waste removed per second, i.e., about 6,700 items of waste removed per hour. Despite this intensive workload, many products could not be picked out by the workers. As a result, the quality targets required by the recyclers for the outgoing waste could not be reached. Consequently, the site management decided to hire an additional employee and to reduce the hourly production (from 7 to 5 metric tons) to limit the workload of each worker and to try to provide the high quality required for the final sorted product. The new organization solved the quality problems. However, it may not be sustainable in the long-term according to the production managers: the reduction in productivity and the increased number of employees, significantly increased production costs.

This example illustrates the challenge of working simultaneously on both the organizational and the technical dimensions. Organizational adjustments are not sufficient to deliver sustainable efficiency for the system and for work-related prevention. Thus, the possibility of adapting the technical equipment to the characteristics of the territorialized waste can be considered with adjustable functions. In this example, this is represented by the inclination of the ballistic separators that can be adjusted through a computer interface by the production managers. Adding this function to the machine represents an additional cost at the design stage, but the function could then cope with predictable and unpredictable changes in the territory and in the MRF.

The coupling between the work systems and the territorialized waste is therefore a plasticity issue. It marks a sort of boundary beyond which the cost of performance becomes too high for the organization and for the health of those who work there. Our findings lead us to make recommendations at the design stage, as described above, to support systems plasticity that is sufficiently tolerant to changes in the territory and in the MRF. However, the analyses also show that, once the choices related to the characteristics of the territory are crystallized at the design stage, not all desired adjustments are possible. Certainly, it is necessary to avoid *hypertely* (in the sense of Simondon, 1958/1989), or going beyond what is useful, in the search for plasticity. Plasticity is not unlimited, except when large investments are allocated for alterations. For their activity in the MRFs, the workers have room for maneuver, i.e., a range of resources to adapt and adjust, and they develop adaptive strategies to cope with the territoriality of the waste, thereby delivering recycling performance without affecting their health. However, the more specialized the MRF (and threatened with *hypertely*), the less adjustments are possible. The plasticity of the work system is thus a dimension that needs to be objectified and supported by cooperation between the actors inside the sorting company and those outside the company, positioned at the level of the territory, for the EPCI, as we will see in the following part.

13.5.3 Cooperation between stakeholders at the different decision-making levels

Cooperation between stakeholders inside and outside the company appears to be a necessity to support a coupling between the MRF and its territory and to improve the efficiency of production system design and work-related prevention. In this research, we chose to work toward better cooperation between the decision-makers from the territorial governance and the managers of the MRF (Boudra, Delecroix, & Béguin, 2015) to act simultaneously on both the technical and the organizational dimensions of the MRFs.

In each MRF, after the work analyses, we organized meetings as "debating spaces" bringing together the site production managers and the local authorities from the EPCI, who were external to the sorting company but whose decisions at territorial level influence the sorting work activity. These meetings were organized and moderated by the researcher-interventionist. The meetings had two aims: (i) to discuss technical and organizational improvements and (ii) to transform the representations of the decision-makers relating to the sorting work itself to guide future design choices through the recommendations for technical-organizational plasticity. The topics discussed were related to the production difficulties encountered on the site during our analysis period and their consequences on the work activity and on the health and safety of employees. Indeed, the effort to achieve cooperation between stakeholders at different decision-making levels fundamentally requires such discussions. Without cooperation, the work activity of the sorting employees becomes merely a means of regulating the difficulties of cooperation at the decision-making level and compromises the possibilities available to them to protect their health. Reciprocally, improving the cooperation between the various stakeholders can be a lever for improving working conditions.

With this aim in view, prevention in connection should be approached and understood as a dimension that is intrinsic to territorialized industrial projects, and that is tied in with the technical and organizational choices relating to the production model.

13.6 DISCUSSION

Design is one of the paths for the development of work-related prevention. Conventionally, these two levels do not often coincide in companies. Consequently, occupational safety and health aspects are often incorporated after technical choices have been specified (Lamonde, Beaufort, & Richard, 2004). Moreover, current design models tend to separate design and implementation (Duarte, 2002), whereas a more holistic approach would tend to promote more efficient prevention (Vilela et al., 2007). Thus, attempting to build prevention actions without investigating the design – and *vice versa* – could significantly limit the scope for action and the efficiency of the systems, since technical objects and how they are integrated into work environments (automation, robotization, etc.) transform the work activity of employees in socio-productive organizations. From this perspective, prevention and technical system design should be considered as processes that benefit the workers, for whom they can offer many resources for action.

In the case of MRFs, specifying the dimensions of technical-organizational plasticity illustrates this possible convergence between design and prevention. But the realities of the work and of production (foreseeable or random variabilities, etc.) remain difficult to be taken into consideration both in the technical choices and in the criteria used to assess industrial performance. As seen in the findings presented in this chapter, the commercial logic that consists in making waste an economic resource tends to steer design choices. Finally, the challenges and issues taken up by decision-makers strongly influence the design choices and define the boundaries of technical-organizational plasticity. Nevertheless, industrial performance cannot be achieved without the adaptive strategies implemented by workers as they perform their activity, sometimes by jeopardizing their health and safety.

In this context, the role of the researcher-interventionist was therefore to take part in transforming the points of view of the decision-makers. This aims to identify what the work requires for the employees (in terms of constraints, duties, and exposure to risk factors), and to identify the resources needed for the activity. This perspective leads to no longer considering that efficiency can be found by increasing the reliability of the technical systems, but rather through the workers' activity, provided that this activity can be deployed in a technical and organizational system that is sufficiently plastic.

However, this perspective assumes a possible meeting between the designers and the users (Béguin, 2009). Such a confluence can be characterized as a "mutual learning process" (Béguin, 2003) that fits into the dynamics of the design project. The meeting may be direct, often at the initiative of the members of the design project (engineers, ergonomists, etc.). It may also be indirect, instigated by decision-makers, as we observed during this research. And some studies argue that supporting a diversification of the types of meetings should facilitate exchanges of views and experiences between users and designers (Darses & Wolff, 2006). Indeed, users' participation in the design project makes it possible to shift the boundaries between design, implementation, and use (Béguin, 2007b). In fact, this topic is a current issue in the field of design and in the field of occupational risks prevention. Participation is often seen as a necessity to design efficient situations that offer developmental possibilities for both the individuals and the systems (Barcellini, Van Belleghem, & Daniellou, 2014; Béguin, 2010; Folcher, *op.cit.*).

Nevertheless, it must be admitted that there are obstacles to building such confluence. These obstacles usually lead to a lack of meeting between the *professional worlds* of designers, decision-makers, and users. Undoubtedly, this lack of meeting can be explained to some extent by the legacy of scientific management (Taylorism), that is still very present in the world of work, formalizing a strict vertical division of labor (Owen, 2009). That would thus suggest that such confluence is dependent on the organizational forms in which the industrial design projects fit. Consequently, this is an issue related to simultaneously identifying design and cooperation criteria (Lamonde, 1996).

Furthermore, a possible meeting between designers and users is dependent on the political choices made by decision-makers; on which designers and users have little or no direct influence. The term "decision-makers" is used herein to mean instructing parties, commissioning clients, funding providers, etc., they do not constitute a homogeneous "professional world" (e.g., Béguin, 2010). For the waste sorting sector in France, the commissioning clients are often members of territorial governance. Local authorities award public contracts to design offices, to design workspace, and technical systems. In this context, the technical and political choices can express imperfect and inaccurate representations of the work to be done and the workers' activities. This is what we can conclude from the findings on the design of multiple sorting cabins in the MRF, or on the inappropriate equipment (notably the mobile disposal equipment).

From this perspective, we proposed technical-organizational plasticity, considering the uses and the human activity *in situ*. On a methodological level, findings from the work analyses become a useful tool for the design and work-related prevention. The action of the ergonomist then aimed to influence the choices of the decision-makers by integrating the real work and health issues to a greater extent. For example, as shown by previous research, considering the characteristics of older populations at work and the development of experience with age (Pueyo & Laville, 1997) are relevant criteria when proposing models for the design of sustainable work systems (Béguin, 2011).

These results encourage us to consider new methods of action for work-related prevention by asserting a need for strong links between the design project and the prevention project: by considering their deployment as the result of collective activities distributed in time and space, made up of several partners and competencies from different professional worlds, contributing to the efficiency of the work system and to the development of the activity, safety, and health of the workers.

NOTES

1 The French Labor Code (*Code du Travail*) stipulates that the employer should take the necessary steps to secure the safety and to protect the physical and mental health of the workers (Art. L.4121.1).

2 *Établissement Public de Coopération Intercommunale.*
3 Such methodological tools are traditionally used in ergonomics paradigm. They are used for reflecting activities, performed with individuals or groups. Researchers present to workers data collected during the observations phase (notes, images, or video) to comment their work activity (constraints, resources built or needed and effects on safety, health, and system) and to trace the possibility to transform the work situation. This framework aims to share work experience, support activity development and redesign work system.
4 The sorting cabin is a room where manual sorting is performed.
5 This information suggests that such strategies are mostly "incorporated" by the workers and not easily verbally accessible. To transmit the knowledge on his or her work activity, the worker needs to replicate the usual movements performed during the interview.
6 A fixed magnetic system that is situated inside a frame around which a removal belt turns, driven by a motor-and-gearbox unit or by a drive pulley, the system is placed above a conveyor belt and allows sorting of ferromagnetic particles and objects.
7 Sound can also be used to detect the presence of items of recyclable or undesirable waste that are not immediately visible because they might be hidden by other waste. For example, in MRF1, flexible and thin plastics were often masked by thick layers of paper on the sorting belt. Workers rubbed piles of paper between their hands to detect whether plastics were present in the piles based on the noise generated. This strategy was used specifically for waste from bagged collection channels. These bags had a smaller volume than the containers. This made it more likely that residents who filled them at home would group different types of waste materials together to limit the volume of the bag stored in their house.
8 When geographical areas are long distances away from the MRF, the waste is temporarily stored in transfer centers, and delivered to the MRF on a weekly basis by semi-trailer trucks.
9 Multi-material collections are composed of packaging and paper, newspapers, and magazines mixed in the same collection.

REFERENCES

Avila-Assunção, A. (1998). *De la déficience à la gestion collective du travail: les troubles musculo-squelettiques dans la restauration collective* (PhD thesis). Paris: EPHE.

Barcellini, F., Van Belleghem, L., & Daniellou, F. (2014). Design projects as opportunities for the development of activities. In P. Falzon (Ed.), *Constructive Ergonomics* (pp. 150–163). Boca Raton, FL: Taylor and Francis.

Béguin, P. (2003). Design as a mutual learning process between users and designers. *Interacting with Computers*, 15(5), pp. 709–730.

Béguin, P. (2007a). In search of a unit analysis for designing instruments. *Artifact*, 1(1), pp. 11–16.

Béguin, P. (2007b). Taking activity into account during the design process. *Activités*, 4(2), pp. 116–121.

Béguin, P. (2009). When users and designers meet each other in the design process. In C. Owen, P. Béguin and G. Wackers (Eds.), *Risky Work Environments: Reappraising Human Work Within Fallible Systems* (pp. 153–172). Surrey: Ashgate Publishing Limited.

Béguin, P. (2010). *Conduite de projet et fabrication collective du travail, une approche développementale* (Professorial thesis). Bordeaux, France: Université Victor Segalen Bordeaux 2.

Béguin, P. (2011). Acting within the boundaries of work systems development. *Human Factors and Ergonomics in Manufacturing & Service Industries*, 21(6), pp. 543–554.

Béguin, P., Owen, C., & Wackers, G. (2009). Introduction: shifting the focus to human work within complex socio-technical systems. In C. Owen, P. Béguin & G. Wackers (Eds.), *Risky Work Environments: Reappraising Human Work Within Fallible Systems* (pp. 1–10). Surrey: Ashgate Publishing Limited.

Béguin, P., & Rabardel, P. (2000). Designing for instrument-mediated activity. *Scandinavian Journal of Information Systems*, 12(1), pp. 173–190.

Boudra, L. (2016). *Durabilité du travail et prévention en adhérence. Le cas de la dimension territoriale des déchets dans l'activité de tri des emballages ménagers* (PhD thesis). Lyon: Université Lumière Lyon 2, France.

Boudra, L., Béguin, P., Delecroix, B., & Pueyo, V. (2019). Prendre en compte le territoire dans la prévention des risques professionnels. Le cas du travail de tri des emballages ménagers. *Le Travail Humain*, 82(2), pp. 99–128.

Boudra, L., Delecroix, B., & Béguin, P. (2015). Taking into account the territorial dimension of work for sustainable work system. The case of waste sorting centres. *19th Congress of the International Ergonomics Association*. Melbourne, August 9th–14th, 2015.

Boudra, L., Pueyo V., & Béguin, P. (2018). The territorial anchorage of waste sorting activities and its organization for prevention. *Proceedings of the 20th Congress of the International Ergonomics Association*. Florence, Italy, August 26th–30th.

Carayon, P. (2006). Human factors of sociotechnical systems. *Applied Ergonomics*, 37(4), pp. 525–535.

Caroly, S., Coutarel, F., Landry, A., & Mary-Cheray, I. (2010). Sustainable MSD prevention: management for continuous improvement between prevention and production. Ergonomic intervention in two assembly line companies. *Applied Ergonomics*, 41(4), pp. 591–599.

Daniellou, F. (2005). The French-speaking ergonomists' approach to work activity: cross-influences of field intervention and conceptual models. *Theoretical Issues in Ergonomics Science*, 6(5), pp. 409–427.

Daniellou, F., Escouteloup, J. & Beaujouan, J. (2011). Phasage des travaux et organisations transitoires: quels rôles pour l'ergonome? *Activités*, 8(1), pp. 1–19.

Darses, F. & Wolff, M. (2006). How do designers represent themselves the users' needs? *Applied Ergonomics*, 37(6), pp. 757–764.

de Keyser, V. (1991). Work analysis in French language ergonomics: origins and current research trends. *Ergonomics*, 34(6), pp. 653–669.

Docherty, P., Forslin, J., & Shani, A.B. (2002). *Creating Sustainable Work Systems*. London: Routledge.

Duarte, F. (2002). Complementaridade entre ergonomia e engenharia em projetos industriais. In Duarte, F. (Eds.), *Ergonomia & Projeto na indústria de processo contínuo* (pp. 11–21). Rio de Janeiro: Lucerna.

Engkvist, I.-L., (2010). Working conditions at recycling centres in Sweden – Physical and psychosocial work environment. *Applied Ergonomics*, 41, pp. 347–354.

Folcher, V. (2003). Appropriating artifacts as instruments: when design-for-use meets design-in-use. *Interacting with Computers*, 15(5), pp. 648–663.

INRS. (2011). *Centre de tri de déchets recyclables secs ménagers et assimilés issus des collectes séparées – Guide de prévention pour la conception. ED 6098.* Vandoeuvre: Institut National de Recherche et de Sécurité.

Kerval, A. (1990). La genèse du mode industriel dégradé. *Le Travail Humain*, 53(4), pp. 369–372.

Lamonde, F. (1996). Safety improvement in railways: which criteria for coordination at a distance design? *International Journal of Industrial Ergonomics*, 17(6), pp. 481–497.

Lamonde, F., Beaufort, P., & Richard, J.-G. (2004). Ergonomes et préventionnistes: étude d'une pratique de collaboration dans le cadre d'un projet de conception d'une usine – 1ère de 2 parties. *Perspectives interdisciplinaires sur le travail et la santé*, 6(1), pp. 1–20.

Lavoie, J., & Alie, R. (1997). Determining the characteristics to be considered from a worker health and safety standpoint in household waste sorting and composting plants. *Annals of Agricultural and Environmental Medicine*, 4, pp. 123–128.

Malmros, P., Sigsgaard, T., & Bach, B. (1992). Occupational health problems due to garbage sorting. *Waste Management & Research*, 10(3), pp. 227–234.

Marc, J., & Rogalski, J. (2009). How do individual operators contribute to the reliability of collective activity? A French medical emergency centre. In, C. Owen, P. Béguin & G. Wackers (Eds.), *Risky Work Environments: Reappraising Human Work Within Fallible Systems* (pp. 129–148). Surrey: Ashgate Publishing Limited.

Mollo, V. & Falzon, P. (2004). Auto-and allo-confrontation as tools for reflective activities. *Applied Ergonomics*, 35(6), pp. 531–540.

Nascimento, A, & Falzon, P. (2012). Producing effective treatment, enhancing safety: medical physicists' strategies to ensure quality in radiotherapy. *Applied Ergonomics*, 43(4), pp. 777–784.

Owen, C. (2009). Accomplishing reliability with-in fallible systems. In C. Owen, P. Béguin and G. Wackers (Eds.). *Risky Work Environments: Reappraising Human Work Within Fallible Systems* (pp. 99–103). Surrey: Ashgate Publishing Limited.

Pearce, D. (1988). Economics, equity and sustainable development. *Futures*, 20(6), pp. 598–605.

Poulsen, O.M., Breum, N.O., Ebbehøj, N., Hansen, Å.M., Ivens, U.I., Lelieveld van, D., Malmros, P., Matthiasen, L., Nielsen, B.H., Nielsen, E.M., Schibye, B., Skov, T., Stenbaek, E.I., & Wilkins, K.C. (1995). Sorting and recycling of domestic waste. Review of occupational health problems and their possible causes. *The Science of the Total Environment*, 168, pp. 33–56.

Pueyo, V., & Laville, A. (1997). Role of experience in a cold rolling mill quality control. *Arbete & Halsa*, 29, pp. 257–262.

Pueyo, V., & Zara-Meylan, V. (2014). Gérer collectivement les situations dangereuses: un exemple dans les hauts-fourneaux. In A. Thébaud-Mony, P. Davezies, L. Vogel & S. Volkoff (Dir.). *Les risques du travail. Pour ne pas perdre sa vie à la gagner* (pp. 167–170). Paris, France: La Découverte.

Pueyo, V., Toupin, C., & Volkoff, S. (2011). The role of experience in night work: lessons from two ergonomic studies. *Applied Ergonomics*, 42(2), pp. 251–255.

Rabardel, P., & Béguin, P. (2005). Instrument mediated activity: from subject development to anthropocentric design. *Theoretical Issues in Ergonomics Science*, 6(5), pp. 429–461.

Rabardel, P., & Pastré, P. (2005). *Modèles du sujet pour la conception: dialectiques, activités, développement*. Toulouse, France: Octarès Éditions.

Rushton, L. (2003). Health hazards and waste management. *British Medical Bulletin*, 68, pp. 183–197.

Schlosser, O., Déportes, I.Z., Facon, B. & Fromont, E. (2015). Extension of the sorting instructions for household plastic packaging and changes in exposure to bioaerosols at materials recovery facilities. *Waste Management*, 45, pp. 47–55.

Simondon, G. (1958/1989). *Du mode d'existence des objets techniques*. Paris, France: Aubier Philosophie.

Sperandio, J.C. (1971). Variation of operator's strategies and regulating effects on workload. *Ergonomics*, 14(5), pp. 571–577.

Suchman, L. (1987). *Plans and Situated Actions*. Cambridge: Cambridge University Press.

Teiger, C. (2007). De l'irruption de l'intervention dans la recherche en ergonomie. *Education Permanente*, 170, pp. 35–49.

Vicente, K. J. (1999). Cognitive work analysis. *Toward Safe, Productive and Healthy Computer-Based Work*. Boca Raton, FL: Routledge, Taylor & Francis.

Vilela, R. A. G, Mendes, R.W.B, & Gonçalvez, C.A.H. (2007). Acidente do trabalho investigado pelo CEREST Piracicaba: confrontando a abordagem tradicional da segurança do trabalho. *Revista Brasileira de Saúde Ocupacional*, 32(115), pp. 29–40.

Visser, W. (2007). Designing as construction of representations: a dynamic viewpoint in cognitive design research. *Human-Computer Interaction*, 21(1), pp. 103–152.

Wisner, A. (1995). Understanding problem building: ergonomic work analysis. *Ergonomics*, 38(3), pp. 595–605.

Chapter 14

Conclusion

Edwige Quillerou
INRS, French Research and Safety Institute for the Prevention of
Occupational Accidents and Diseases
Vandœuvre-lès-Nancy, France

Guy André Boy
Paris Saclay University (Centalesupélec) and ESTIA
Institute of Technology
Gif-sur-Yvette & Bidart, France

This book provides an account of cross-fertilization of prevention and design for risk management. Risk-taking is introduced as a necessity in many situations where procedural knowledge is not applicable or is incomplete, and where problem-solving leads to taking risky actions. This issue leads to the analysis of risk in concrete real-world situations and prevention design. It is perfectly natural to take risks, and we cannot perceive and understand risk until we have seen it up in practice. Risk-taking is about experiencing risk, which is challenging, since it leads to exposing people to danger. Testing early enough during the design and development of a system is often mandatory to simulate the risk experience. How far can we go when testing in risky situations? The answer to this question is empirical and based on specific accumulated and compiled expertise.

Accident analysis can be very informative and have an impact on the design of new systems. In other words, safety history and therefore culture are precious for the development of novel technology, organizations and related people's activities. However, observation of what happened is not enough: methods that enable us to analyze risk in sociotechnical systems are needed. It is commonly observed that experience feedback is based on human error analysis of the negative experience. Conversely, feedback of positive experience is also required to relate these to the achievement of successful endeavors. Group dynamics and creativity are key assets for a better shared situation awareness of how the system works and how it can be controlled and managed. Human-in-the-loop simulation can be extremely useful for observing and analyzing social networks. Sociological and anthropological investigations are always crucial for capturing concrete real-world network structures and functional interconnectivity among the agents involved.

As it is necessary to provide real-world experiences in safety-critical sociotechnical design, human systems integrators must investigate

DOI: 10.1201/9781003221609-14

situation awareness, decision-making and action taking to anticipate possible futures. In high-level sports, for example, preparation is the key to build automatisms and problem-solving skills that will be mandatory at performance time, purposefully, accurately and timely. This can be applied in other domains such as road maintenance. Submarine experience informs us that a group of people isolated in a small and confined environment needs to develop collaborative skills and knowledge for longer sustainable survival. This brings back the generic need for preparation in risk-taking experience (Boy & Brachet, 2010).

Risk mitigation has been described in commercial aviation and submarine operations as examples. This is commonly considered in the development of operations procedures, but even more in human-centered design. Following up these use cases, both airline pilots and submariners are now unavoidable contributors in participatory design setups. They even can become designers themselves. No matter what, the future of engineering design and systems engineering will necessarily involve operations people. In other words, doers should be associated with thinkers. However, life is not a steady state for a static class of people. It should be dynamic because people not only evolve in competence but also with age. Aging is a factor that is required to be considered in risk-taking. Indeed, people are likely to become wiser, even if their reaction skills may erode with time.

We have seen that cross-fertilization of prevention and human systems integration is useful for human-centered and organizational design. A use case was presented in the medical sector, where physicians often must take risks to try possible actions to solve problems that cannot be well stated for time pressure constraints in emergency situations for example. Critical care and reanimation personnel are required to act in a hurry to save lives; these actions should be based on specific training and accumulated expertise. Procedures are not enough in emergency situations, and creativity is often required. At this point, organization design and management enter into play. This is the case in pandemic situations, such as COVID-19, where no procedure is capable of supporting physicians, and empirical problem-solving should take place – another case where zero risk does not exist! Again, doing nothing because conventional and legal regulations block such problem-solving can be extremely dangerous and induce other risks.

Anticipation of risk-taking can be considered at design time. At this point, collaboration between various kinds of experts is necessary. In the OWSAS case, we have seen that meteorologists, aeronautical engineers, pilots and human activity specialists (i.e., preventers) have to work together to end up with a satisfactory solution. Participatory design that involves these experts is more likely to lead to optimize risk-taking and management. In addition to depth-first design of a local system (e.g., OWSAS), breadth-first design of a global system (e.g., waste management and recycling system) connects the various agents and systems involved. We usually talk about systems of systems.

Summing up, cross-fertilization of engineering design and human prevention is a matter of connecting technology, organizations and people during the whole life cycle of a system of systems. This is not possible without considering and implementing activity observation and analysis. In this book, activity relates to Leontiev's human activity theory (Leontiev, 1978).

What are the perspectives? The role of science and the way of doing science are changing. We should plan on doing science in action (i.e., the learning-by-doing side of science), and sometimes we cannot avoid taking risks. In other words, managers should learn how to support creativity of operational contributors and face risks in a safer and exploratory way. This does not go without appropriate resources and reasonable flexibility. This is because our world is becoming more complex every day with the development of digital technology and increasing connectivity. We live inside virtual environments where sociomedia plays a crucial role. This is the reason why tangibility has become so important. Tangibility is not only physical (i.e., the relationship with physics and mechanics) but also cognitive and figurative (i.e., the meaning of things we interact with) (Boy, 2016, 2020). Of course, these two facets of tangibility are related inside human beings, where emotions, reasoning, concentration and sensing the world around us are intimately related.

In engineering design and systems engineering, uncertainty is always present at any point in time. If uncertainty cannot be avoided, it should be constantly managed. Uncertainty management in life-critical systems (Boy, 2021) requires good experience feedback both short term and longer term. Indeed, operational personnel know better than anyone else about the domain at stake (Quillerou & Lux, 2019). Therefore, their knowledge should be used to manage uncertainty. We cannot predict the future, but we can invent it[1]... and test it in modeling and simulation.

These issues of science in action and uncertainty management force us to constantly redefine scientific epistemology. François Rabelais said, "Science without conscience is only ruin of the soul". This is crucial for human development, in Vygotsky's sense of the development of human capacities built in interaction with other people, and we are accountable to these other people. Everybody has the power of action.

Problem is that risks are omnipresent in our daily life at various levels of danger, especially at work. Whether in industry, defense, medicine, aviation, sports and so on, safety-critical systems have common grounds that are omnipresent and need to be investigated during the whole life cycle of sociotechnical systems (i.e., from design to dismantling). Safety culture is incrementally built through endless collaboration of engineering designers and preventers. What is designed is not only technology, as conventional engineering used to do, but also concurrently organizations and jobs. It takes time to get to mature states at work because new know-how and properties always emerge and need to be eventually incorporated into current practice. Indeed, even if we start to have a maturing technology, there are always emerging properties being discovered. These emergent properties contribute to changing practices in a longer term domain culture. For example, the car industry matured during the 20th century both technologically and with respect to various practices, such as transporting goods leading to the development of the truck category that is very different from the racing car category. Culturally, people structured their life around automotive mobility.

Designing for large groups of people and a variety of operations requires elicitation of appropriate attributes and criteria, such as safety, efficiency and comfort. In the beginning of a design project, considering these attributes and criteria up-to-front is always difficult and leads to uncertainty management. Anthropological studies should be carried out to better understand how people deal with new technology based on human-in-the-loop simulations. People are always guided in their activity by cognitive and emotional motivations (Norman, 1988), and local work practices (Wisner, 1997). Uncertainty and system knowledge is incrementally optimized by endless testing of technology being developed. This kind of process cannot be done without a tight collaboration between engineering designers and social science experts. This is the reason why human systems integration (HSI) becomes the result of the cross-fertilization of these backgrounds within the human-centered design (HCD) approach, integrating individual and collective dimensions. In addition, risk-taking and risk management involve both complexity analysis and familiarization with complexity through the combination of various competencies and viewpoints. This research-by-doing approach leads to the creation of novel competencies and organizational setups.

Risk at work is multi-factorial. This book presented various types of risk in a large variety of situations and domains. It is now crucial to consider risk as upstream as possible in the life cycle of a sociotechnical system. In this way, we will be able to improve prevention, instead of developing mitigation and recovery procedures as it is typically the case today. This is the reason why participatory design is so important, where preventors, designers and end users should work together to incrementally come up with appropriate solutions. This book gathered this kind of variety and can be a springboard toward the cross-fertilization, which was our target. The resulting HSI developmental approach should be scaled up from the operational level to the strategic level. The recent memorandum of understanding between INCOSE (International Council on Systems Engineering) and IEA (International Ergonomics Association) at the world level is a historical step toward the development of transdisciplinary and complementary teams. I would like this book to be an initial support of future research and industrial collaborations.

NOTE

1 This quote has been referred to many people, including Dennis Gabor, Abraham Lincoln, Ilya Prigogine, Alan Kay, Steven Lisberger, Peter Drucker and Forrest C. Shaklee.

REFERENCES

Boy, G.A. (2016). Tangible Interactive Systems. Springer, London, UK.

Boy, G.A. (2020). Human Systems Integration: From Virtual to Tangible. CRC Press – Taylor & Francis Group, Boca Raton, FL.

Boy, G.A. (2022 to appear). Managing Uncertainty in Life-Critical Systems. In The Handbook of Uncertainty Management in Work Organizations, G. Grote & M. Griffin (Eds.). Oxford University Press, Oxford, UK.

Boy, G.A. & Brachet, G. (2010). Risk Taking: A Human Necessity that Needs to be Managed. Air and Space Academy, Toulouse, France.

Bruner, J. (1996). *The Culture of Education*. Cambridge, MA: Harvard University Press.

Gaba, D. M., Maxwell, M., & DeAnda, A. (1987). Anesthetic mishaps: Breaking the chain of accident evolution. *Anesthesiology clinics*, 66(5), 670–676.

Leontiev, A. (1978). Activity, Consciousness, and Personality. http://www.marxists.org/archive/leontev/works/1978/index.htm.

Norman, D.A. (1988). The Design of Everyday Things. Basic Books, New York.

Quillerou, E. & Lux, A. (2019). "Orchestrating" Methodology for Designing an Industrial System that Improves Occupational Health and Safety. Human Systems Integration Conference (HSI2019), Biarritz, France, September 11–13, 2019. https://www.incose.org/hsi2019/conference/event-schedule.

Vygotski, L. S. (1980). *Mind in society: The development of higher psychological processes*. Cambridge, Massachusetts & London, England: Harvard university press.

Woods, D. (2019). Essentials of resilience, revisited. In M. Ruth & S. Goessling-Reisemann (Eds.), *Handbook on Resilience of Socio-Technical Systems* (pp. 52–65). Northampton: Edward Elgar Publishing.

Woods, D., & Dekker, S. (2000). Anticipating the effects of technological change: a new era of dynamics for human factors. *Theoretical Issues in Ergonomics Science*, 1(3), 272–282.

Wisner, A. (1997). Anthropotechnology: Towards a Pluricentric Industrial World. In French, Anthropotechnologie: Vers un monde industriel pluricentrique. Octares, Toulouse, France.

Index

Milton Keynes UK
Ingram Content Group UK Ltd.
UKHW031147141024
449569UK00024B/1012